Living with Animals: a Zooarchaeological Study of Urban Human-Animal Relationships in Early Modern Tornio, 1621-1800

Anna-Kaisa Puputti

BAR International Series 2100
2010

Published in 2016 by
BAR Publishing, Oxford

BAR International Series 2100

Living with Animals: a Zooarchaeological Study of Urban Human-Animal Relationships in Early Modern Tornio, 1621-1800

ISBN 978 1 4073 0576 9

© A-K Puputti and the Publisher 2010

The author's moral rights under the 1988 UK Copyright,
Designs and Patents Act are hereby expressly asserted.

All rights reserved. No part of this work may be copied, reproduced, stored, sold, distributed, scanned, saved in any form of digital format or transmitted in any form digitally, without the written permission of the Publisher.

BAR Publishing is the trading name of British Archaeological Reports (Oxford) Ltd. British Archaeological Reports was first incorporated in 1974 to publish the BAR Series, International and British. In 1992 Hadrian Books Ltd became part of the BAR group. This volume was originally published by Archaeopress in conjunction with British Archaeological Reports (Oxford) Ltd / Hadrian Books Ltd, the Series principal publisher, in 2010. This present volume is published by BAR Publishing, 2016.

Printed in England

PUBLISHING

BAR titles are available from:

         BAR Publishing
         122 Banbury Rd, Oxford, OX2 7BP, UK
EMAIL   info@barpublishing.com
PHONE   +44 (0)1865 310431
  FAX   +44 (0)1865 316916
         www.barpublishing.com

Contents

List of figures
List of tables
List of appendices
Acknowldegements
1. Introduction — 1
    1.1. The aims and outline of the study — 1
    1.2. The town of Tornio 1621-1800 AD — 1
    1.3. Livelihood in Tornio — 3
2. Theoretical framework — 4
    2.1. Human-animal relationships — 4
    2.3. Towards a zooarchaeology of people — 4
3. Zooarchaeological methods — 7
    3.1. Description of the individuals — 7
    3.2. Quantitative methods — 8
4. Archaeological material — 10
    4.1. Archaeological sites — 10
    4.2. Overview of the bone material — 11
    4.3 Disposal of animal remains — 12
5. Analysis of the bone material — 14
    5.1. Patterns in species diversity — 14
    5.2. Animal husbandry practices — 17
        Cattle — 17
        Sheep or goat — 25
        Pig — 30
        Reindeer — 32
        Horse — 34
        Domestic hen — 34
    5.3. Hunting wild species — 34
        Seals — 34
        Arctic hare — 34
        Red squirrel — 35
        Red fox — 35
        Brown bear — 35
        Cervids — 35
        Gallinaceous birds — 35
        Waterfowl — 36
6. Discussion — 37
    6.1. Diet differences, social status and identity — 37
    6.2. Living with animals — 38
    6.3. Attitudes towards wild animals and wilderness in seventeenth-century Tornio — 40
    6.4. Modernising human-animal relationships — 40
7. Conclusions — 43
Appendices — 44
References — 67
Electronic sources — 74
Archive sources — 74

List of figures

Figure 1. a) The location of Tornio. (Drawing A. Puputti) b) Map of Tornio in 1647 by Nikodemus Tessin Senior. (Archive source: Ut. Känd proven. Nr, 433 (kartavd.m.form).RA.)   2

Figure 2. The location of Tornio and the excavation areas situated on the map of present-day Tornio. (Drawing A. Puputti)   10

Figure 3. The spatial and temporal distribution of the animal bone material according to weight (kg).   11

Figure 4. The distribution of bones of wild and domestic animals in different context types in the Keskikatu excavations.   13

Figure 5. The distribution of bones of different-sized animals in different context types in the Keskikatu excavations. 13

Figure 6. The proportions of bones of wild animals of all bones identified to species or genus (% NISP) and proportions of cattle bones of all bones of domestic animals (%NISP) in different excavation areas and periods.   15

Figure 7. The proportions of wild animals of all the bone fragments identified to species or genus (% NISP) plotted against the number of wild species encountered in the assemblage.   16

Figure 8. Proportions of bones of wild animals of all animal bones identified to species or genus in different periods, and the number of wild species in each period.   16

Figure 9. Combined data on tooth eruption and wear of cattle.   18

Figure 10. Proportions of fused epiphyses in each age group, divided by period.   18

Figure 11. The scatterplot of the distal breadth and depth (mm) of cattle metacarpals.   19

Figure 12. The boxplot of the maximum depth of fused (N= 14 ± 1.9) and unfused (N = 10 ± 2.6) cattle metacarpals at the distal fusion point (mm).   19

Figure 13. Comparison of log ratios of cattle skeletal measuremens between the periods 1620-1660 (N= 124 ± 0.03), 1650-1728 (N= 48 ± 0.02) and 1721-1800 (N= 43 ± 0.03).   22

Figure 14. A cattle metatarsal with craniocaudally split distal end.   23

Figure 15. Scatterplots of bone mineral densities of wildebeest skeletal elements according to Lam et al. (1999) and minimum numbers of elements (MNE) of cattle in different periods.   24

Figure 16. Proportions of fused epiphyses in each age group, divided by period.   25

Figure 17. Combined data on tooth eruption and wear of ovicaprids.   25

Figure 18. The scatterplot of the physiological distal breadth (measurement definition in Davis 1996) and shaft width (mm) of sheep metacarpals. m? = possible males.   26

Figure 19. The scatterplot of the greatest lenght (mm) and slenderness index of sheep metacarpals. m? = possible males.   26

Figure 20. Comparison of log ratios of sheep or goat skeltal measuremens between the periods 1620-1660 (N= 253 ± 0.04), 1650-1728 (N= 51 ± 0.05) and 1721-1800 (N= 18 ± 0.03).   27

Figure 21. Scatterplot of bone mineral densities of sheep skeletal elements according to Lyman (1994:240-248) and minimum numbers of elements (MNE) of sheep or goat.   29

Figure 22. Combined data on tooth eruption and wear of pigs.   30

Figure 23. Proportions of fused epiphyses in each age group, divided by period.   30

Figure 24. Proportions fused epiphyses in each age group, divided by period.   32

Figure 25. Scatterplot of bone mineral densities according to Lam et al. (1999) and minimum numbers of elements (MNE) of reindeer.   33

List of tables

Table 1. Relative abundances of all bone fragments identified to species or genus (% NISP). 14
Table 2. Withers height estimations of cattle based on metacarpal greatest lenght, estimated with the coefficients of Matolcsi and Fock (von den Driesch & Boessneck 1974). 20
Table 3. Body mass estimations of cattle based on skeletal measurements, according to regression equations of Scott (1990). 21
Table 4. Comparison of withers height and body mass estimations (means, sample sizes and standard deviations) of cattle between periods. 21
Table 5. Bone mineral densities of wildebeest skeletal elements according to Lam et al. (1999) and minimum numbers of elements (MNE) of cattle in different periods. 24
Table 6. Withers height estimations of sheep based on metacarpal greatest lenght, estimated with the coefficient of Teichert (von den Driesch & Boessneck 1974). 28
Table 7. Body mass estimations of sheep or goat based on skeletal measurements, according to regression equations of Scott (1990) and coefficient of O'Connor (2003). 28
Table 8. Bone mineral densities of sheep skeletal elements according to Lyman (1994:240-248) and minimum numbers of elements (MNE) of sheep or goat in different periods. 29
Table 9. Body mass estimations of pigs based on skeletal measurements, according to regression equations of Scott (1990). 31
Table 10. Bone mineral densities of skeletal elements according to Lam et al. (1999) and minimum numbers of elements (MNE) of reindeer in the early seventeenth century. 33

List of appendices

Appendix 1. Numbers of identified specimens (NISP) in each excavation area in 1621-1660. 44
Appendix 2. Numbers of identified specimens (NISP) in each excavation area in 1650-1728. 46
Appendix 3. Numbers of identified specimens (NISP) in each excavation area in 1721-1800. 47
Appendix 4. The skeletal frquencies of wild species as number of identified specimens (NISP), minimum number of elements (MNE) and modified anatomical units (MAU), divided by period. 48
Appendix 5. Age estimations based on cattle mandibulae. Age estimations based on tooth eruption and mandibular wear stages (M.W.S.) are presented. 49
Appendix 6. Sex assessments of cattle based on the pelvis. The minimum thickness of the medial wall of the acetabulum and the morphology of the fossa musculus rectus femoris and pubic bone are presented. 51
Appendix 7. The measurements (mm) of cattle metacarpals and the slenderness index (100*(mc Bd/mc GL)). 53
Appendix 8. The mean, number of specimens, standard deviation and range of postcranial measurements (mm) of cattle, ovicaprids and pigs. 54
Appendix 9. The log ratios of postcranial skeletal measurements (means, sample sizes and standard deviations) of cattle, ovicaprids, pig and reindeer divided by period. 57
Appendix 10. The skeletal frquencies of cattle as number of identified specimens (NISP), minimum number of elements (MNE) and modified anatomical units (MAU), divided by period. 58
Appendix 11. Age estimations based on ovicaprid mandibulae. Age estimations based on tooth eruption and mandibular wear stages (M.W.S.) are presented. 59
Appendix 12. Sex assessments of sheep based on the morphology of the pelvis. 60
Appendix 13. The skeletal frquencies of sheep or goat as number of identified specimens (NISP), minimum number of elements (MNE) and modified anatomical units (MAU), divided by period. 62
Appendix 14. Age estimations based on pig mandibulae. Age estimations based on tooth eruption are presented. 63
Appendix 15. Sex assessments of pig based on the morphology of the canine. 64
Appendix 16. The skeletal frquencies of pig as number of identified specimens (NISP), minimum number of elements (MNE) and modified anatomical units (MAU), divided by period. 65
Appendix 17. The skeletal frquencies of reindeer as number of identified specimens (NISP), minimum number of elements (MNE) and modified anatomical units (MAU), divided by period. 66

Acknowledgements

This book is based on my PhD thesis, completed in 2009 at the University of Oulu. The research was carried out with financial support from the Material Roots of Modernisation in Northern Finland – project and the Finnish Graduate School in Archaeology.

First and foremost, I want to thank my supervisors Vesa-Pekka Herva and Markku Niskanen for incessant help and encouragement during the years I have spent on this study. I would also like to thank my referees Jan Storå, Lembi Lõugas and Kristiina Mannermaa for their valuable comments on the earlier manuscript of this work. I am also most grateful to the historical archaeology and osteology groups at the University of Oulu, the Theoretical Archaeology Discussion Group (TePi) and all our post-graduate students, especially Tiina Äikäs, Tiina Kuokkanen and Sanna Lipponen for inspiring discussion, support and friendship. I also thank Anna Salmi for the language revision of Chapters 1, 2, 6 and 7.

1. Introduction

1.1. The aims and outline of the study

The early modern towns of northern Finland were full of animals: people kept cattle, horses, sheep, goats, pigs, hen, cats and dogs in the town. They hunted and fished in forests, lakes, rivers and the sea and brought their bag or catch back home. Towns and their environs also supported a number of wild birds, mammals and insects. Animals were cared for, pursued, traded, eaten, worked with and processed for soap, glue and hides. Animals were also present in stories, beliefs, rituals, everyday discourse and routines. Animals were an integral part of the urban milieu and the everyday life of people in towns and they cannot be neglected when the urban environments and the worldviews of the past are studied.

This study is an attempt to understand human-animal relationships in the early modern (ca. 1500-1800 AD) town of Tornio in northern Finland with the aid of archaeological animal bone finds from urban excavations. Animal bones from urban contexts can be, for instance, discarded food leftovers, slaughtering waste, buried companion animals, wild animals living and dying in the urban environment or ritual bone deposits (e.g. Armitage 1982; Coy 1982; Stallibrass 2000; Hukantaival 2007). They are remains of living creatures in the towns and they tell us how people treated and shared the bodies of different animal species. Thus, archaeological animal bone finds give a unique insight into the physical presence of animals in the urban milieu and suggest ways in which animals were incorporated in the everyday life and social interaction of people in the past.

The aim of this book is twofold. First, I will describe the animal husbandry practices and the use of wild resources in early modern Tornio based on zooarchaeological evidence. The animal bone assemblages from Tornio have not previously been published or reported, and the urban animal husbandry practices and the use of wild resources have not been analysed archaeologically, apart from a preliminary analysis of the seventeenth-century faunal materials from two plots (Puputti 2006a; b). Second, I will use these results to consider the connections between animals and urban social interaction, and the changing human-animal and human-environmental relationships in early modern Tornio. In this sense, the study also contributes to the understanding of the emerging modern worldview and social order in the northern European periphery during the early modern period. I seek to answer the following questions:

1) Which were the subsistence strategies in animal husbandry and hunting in early modern Tornio? Which animals were kept or hunted, and what resources were gained from these animals? What was the role of wild resources in early modern urban subsistence?

2) Are there dietary or subsistence-related differences within the town? Can these differences be linked to the subsistence strategies or dietary preferences of different urban social groups? How do these results relate to social interaction and power relationships in early modern Tornio?

3) What were human-animal relationships like in Tornio? Can people's attitudes to different animals be inferred from the patterns in the discarded animal bone material? What kind of changes occurred in human-animal relationships and how these changes relate to the emerging modern worldview in northern Finland?

1.2. The town of Tornio 1621-1800 AD

Tornio is a small town in northern Finland (Fig. 1), founded in 1621 as a result of the Swedish crown's policy to control the profitable northern trade. The town was founded by a royal decree at a former marketplace, and the people who moved there came from both present-day Finland and Sweden and were of both local and more distant origins (Mäntylä 1971:33-36). Tornio was situated in northern periphery; it was the northernmost town of Europe in the early modern period, and the subsistence and the economy of its inhabitants still relied heavily on the exploitation of wild resources (Mäntylä 1971:49-50; Ranta 1981:53-57; Puputti 2006a). On the other hand, Tornio was connected to a vast network of trade relationships both in northern and southern directions and it was the starting point for new ideological and economical currents in northern Finland (Mäntylä 1971:46-50). Several rescue excavations have taken place in Tornio during the last two decades, although most of them small in scale. At present, several researchers at the University of Oulu have completed studies on the archaeology of Tornio (Nurmi 2004; Pääkkönen 2006; Salo 2007; Ylimaunu 2007), and active research is still ongoing. Thus, there is a considerable body of archaeological data available, especially from the seventeenth century.

The population of Tornio was constantly changing during the first decades after the town was founded, and the situation seems to have stabilised only during the late seventeenth century (Mäntylä 1971:38; 143). Estimating population size and composition from documentary sources is extremely difficult, but it appears that the population fluctuated somewhere around 200 people during the first half of the seventeenth century and rose to a few hundred in the late seventeenth century and the eighteenth century (Mäntylä 1971:36-37, 132-133). Initially, the population consisted mostly of merchants with mixed economy of trade, animal husbandry, farming, fishing and hunting (Mäntylä 1971; Puputti 2006). The number of specialised craftsmen was quite low, and merely ca 10 % of household heads are estimated to be craftsmen (Mäntylä 1971:51). The number of craftsmen did not increase until the eighteenth century (Mäntylä 1971:386-389).

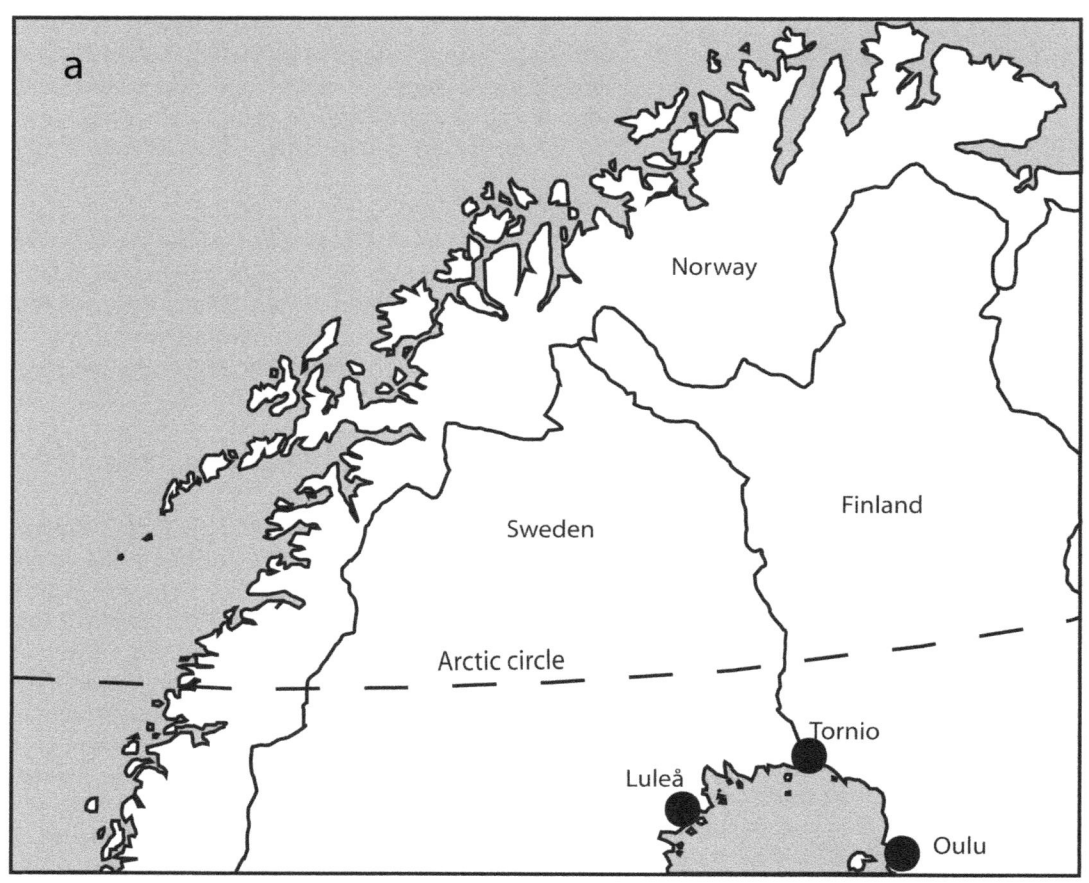
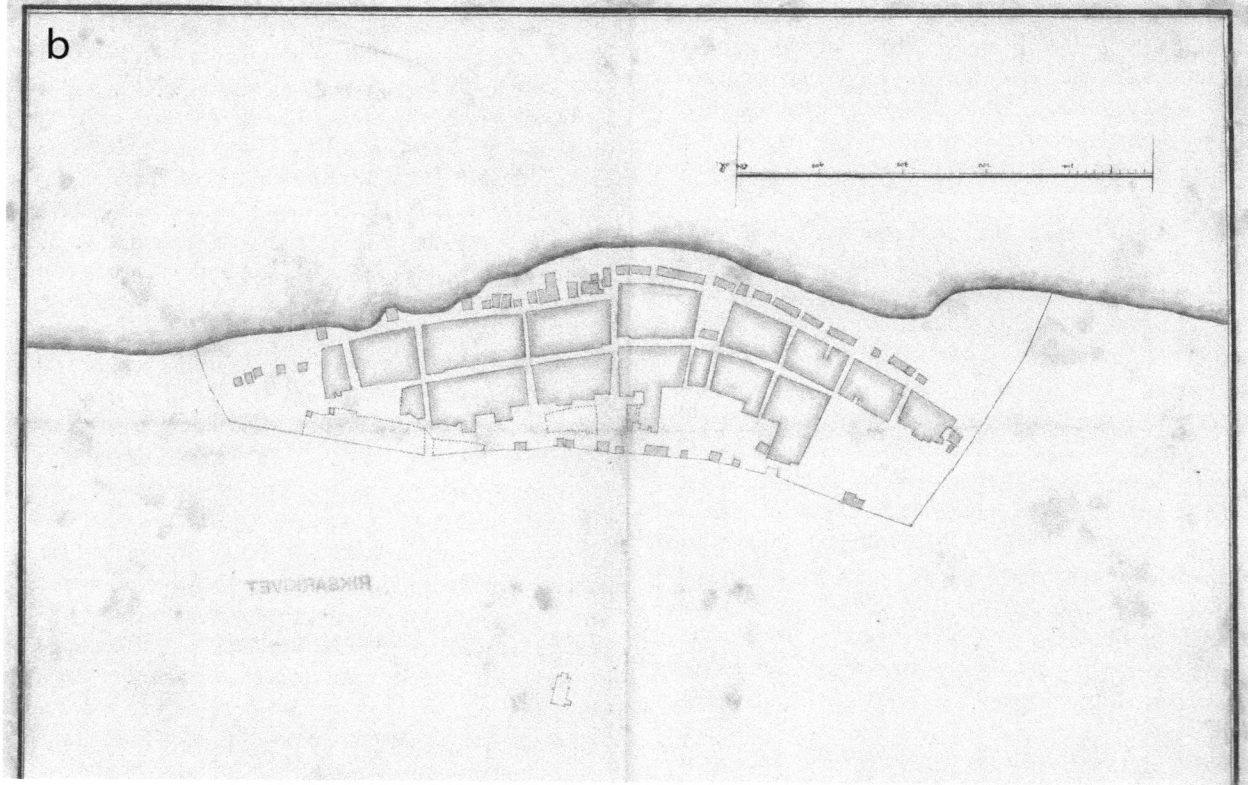

Figure 1. a) The location of Tornio. (Drawing A. Puputti) b) Map of Tornio in 1647 by Nikodemus Tessin Senior. (Archive source: Ut. Känd proven. Nr, 433 (kartavd.m.form).RA.)

Virtually all goods from Lapland were traded via Tornio, at least supposedly, as illegal trade seems to have been common in rural areas (Mäntylä 1993:184). Merchants of Tornio bought fish, butter, reindeer and cattle hides, furs, tar, seal blubber and feathers from Lapland and rural areas of northern Finland, and traded them for salt, grains, cloth, spices, tobacco, malt, hemp and linen with merchants from Stockholm, Baltic region and Russia (Mäntylä 1971:48-50, 70-84). The archaeological material from early modern Tornio is rather pan-European in character. It includes, for instance, clay pipes from England and Holland and ceramics from Central Europe (Nurmi 2004). However, the usage pattern of the artefacts shows a distinct local pattern with constant repair and re-use of materials and artefacts, and an apparent lack of interest in consumption of commercial goods (Salo 2007; Herva & Nurmi 2009).

Wealth was rather unevenly distributed among the townsfolk in Tornio, as the majority of the trade concentrated in the hands of a few rich merchants (Mäntylä 1971:114). Despite its relative wealth, the town remained small and agrarian in appearance throughout the early modern period. It was only during the eighteenth century that its layout and architecture took on a more urban appearance, with increasing stone building technology and enclosed plots (Ylimaunu 2007). Small-scale farming and animal husbandry were also practiced in the town throughout the early modern period. The survival of agrarian and traditional features in the urban landscape can be seen, for instance, in the presence of fields inside the toll fence up to the 1730s (Herva & Nurmi 2009) and animal husbandry within the town (Mäntylä 1971:408). Herva and Nurmi (2009) interpret these phenomena on one hand as an indication of uncertainty about the sustainability of the town, and on the other hand as local resistance to the regulation and modernisation attempts made by the state.

1.3. Livelihood in Tornio

Farming and animal husbandry were an essential part of urban life in the small towns of early modern Finland. Animals, such as cattle, sheep, pigs and hen were commonly kept and people had small vegetable gardens, meadows and fields (e.g. Virkkunen 1953:569; Mäntylä 1971:52, 120). Indeed, it seems that nearly every household in early modern Tornio kept some animals, at least a cow, and the wealthiest people often had several horses, cattle, sheep and even reindeer (Mäntylä 1971:121-123; OMA BIa:1–9.). It has been argued, based on historical documents, that people in early modern Tornio were essentially self-sufficient in farm products such as milk and meat (Virrankoski 1973:249).

Animal husbandry in early modern Tornio seems to differ from that practiced in rural areas. Most importantly, the number of horses in the tax records is considerably higher in the town than in the surrounding countryside, which may be due to the fact that merchants needed them in travelling (Mäntylä 1971:52; Virrankoski 1973:248). In addition, it seems that the ratio of heifers is lower in the town than in the countryside, which may, according to Mäntylä (1971:52), indicate that cattle were bought from the countryside only when they had reached maturity and started to produce milk. The representation of animals in the tax records, unfortunately, is fragmentary and far from complete, since not all animal species (such as domestic fowl, pigs and sheep) were taxed and merchants also owned farms and animals outside the town (Mäntylä 1971:52).

Animals were transported to the western bank of the River Tornionjoki to pasture during the summer. It was forbidden to keep cattle in the town in the summer because they would eat the crops growing in the fields (Mäntylä 1971:121). During the winter, animals were kept in animal shelters in the town. The structure of the animal shelters in early modern Tornio is unclear, but traditionally, sheds with earthen-floored stalls and covered pathway in the middle were used in northern Finland (Vilkuna 1976:72-73; Talve 1990:56). One possible animal shelter was discovered in the Aspio & Viippola excavations (Ylimaunu 2007:40). It was relatively small, ca. two meters in width and of unknown length. The earthen floor was covered with needles and twigs, and remains of timbers in the north-western corner of the building may have been a part of a covered pathway (Ylimaunu 2007:40-41).

Commonly, the horse was the most valuable domestic animal both in terms of price and the attention paid to its care (Virrankoski 1973:240; Soininen 1974:225). Cattle were generally kept for milk production in early modern Finland and mainly old dairy animals were eaten (Virrankoski 1973:239). Pigs were kept solely for meat and generally they were left to take care of themselves and find their own food (Soininen 1974:227; Wilmi 2003:178). Estimating the number of sheep, goats and domestic fowl in early modern Tornio is difficult because these animals were not taxed during the late seventeenth century (Mäntylä 1971:121). However, sheep were generally kept for wool production in rural northern Finland, whereas goats were virtually absent in this area (Alamäki 1956:48; Wilmi 2003:178).

In addition to trade, farming and animal husbandry, also hunting had a significant role in early modern subsistence in Finland. Historical documents concentrate mainly on exported products such as reindeer hides, furs, feathers and seal blubber (Mäntylä 1971:70-71). These were important export products from northern Finland especially during the medieval period and the 16th and seventeenth centuries, while their role in the economy decreased during the eighteenth century (Halila 1953:229-248; Luukko 1954:380-381; Virrankoski 1973:270). However, historical data on home-economic hunting in rural areas indicates that hunting seems to have been quite common in northern Finland (e.g. Teerijoki 1993:93; Vahtola 1997:127). There is no historical data on the hunting activities of townspeople, although animal bone finds from early modern towns indicate that hunting had a considerable role also in urban subsistence.

2. Theoretical framework

2.1. Human-animal relationships

The purely functional and economic way in which people in modern western society and scientific discourse often regard animals and nature has been contested in anthropological theory for some time (Bird-David 1990; Descola & Pálsson 1996; Ellen & Fukui 1996; Viveiros de Castro 1998). The way people perceive and classify animals and natural phenomena, as well as the boundaries drawn between groups of people, people and animals, wild and domesticated spheres of life and edible and non-edible animals are recognised as being culturally constructed (Douglas 1984; Hell 1996; Howell 1996; Knight 1996). The way people consider animals and share their living space with them is dependent on the cultural context, and animals are present in the economic, symbolic, social and practical spheres of human life.

Animals often occupy a prominent role in human thought, religious symbolism and myths, and they can also be powerful symbols in social interaction and social control (Willis 1994; Douglas 1999). Animal fables and myths, and characteristics attached to certain animals, are used to maintain, reproduce, emphasise or challenge existing power relations (Saladin D'Anglure 1994; Salisbury 1994:103-128; Tapper 1988; Sax 2000:47-100). Along the same line, the classification of animals as pure and impure or edible and non-edible reflects cultural boundaries (Douglas 1984). Preferring to eat the meat of certain animals and rejecting others can be used to demarcate belonging to a group and the proper way of life of this group (Douglas 1984:54-57; Salisbury 1994:59-61). Thus, the images and values reflected upon the animal world are closely intertwined with negotiating and maintaining social hierarchies and identities.

In animal symbolism and the social control of food choices, the human society is reflected on the animal world. However, the animal world is not a mere reflection of the sphere of human relations, but the human and animal worlds are organised according to the same principles (Douglas 1994) and animals can be incorporated in the web of social relationships (Ingold 2000:48-52). The boundaries between human and animal, animate and inanimate, social and non-social or wild and domesticated are fluid and culturally constructed (Howell 1996; Knight 1996; Holm 2002). It is acknowledged that often also non-human beings are attributed with personality and regarded as a part of the society (Bird-David 1990; Viveiros de Castro 1998; Ingold 2000:43-52). For instance, the act of hunting requires a reciprocal relationship between the hunter and his prey that includes mutual trust and communication (Brightman 1993:187-189; Ingold 2000:69-72). This kind of relational worldview suggests that animals, humans and forces of nature alike can be described in terms of their involvement with the world, and they can have meaningful interaction with each other (Ingold 2000:43-52).

Such an approach to human-animal relationships emphasises the practical involvement of humans and animals in, for instance, hunting, ploughing, riding or husbanding animals (Ingold 2000:43-57). The relationships between people and animals, as those between people, form and gain their meaning as a result of engaging in mutual interaction (Armstrong Oma 2007:62-76). For instance, the interaction between people and horses in early modern Tornio consisted of shared living space around the same courtyard, feeding and caring for horses, horse-back riding, travelling in a horse carriage and ploughing the fields together with horses, as well the social norms and cultural ideas concerning horses. Similarly, the relationship between a hunter and his prey constitutes of the rituals, stories and beliefs concerning the animal, knowledge of its behaviour, entering the forest, tracking the animal, killing the animal with hunting equipment and the treatment of the remains of the animal. It has been argued (Brück 1999; Ingold 2000:57) that the spiritual and material aspects of hunting are not separated in the mind of the hunter, but are incorporated as something that can be described as 'dwelling' in the environment (Ingold 2000:57) or a culturally constructed rationalisation of human-animal relationships (Brück 1999:322).

Because the relationships between people and animals are constructed in daily encounters and communication, they are closely intertwined with economic activities and subsistence. Changes in subsistence patterns, economy and interaction with different animal species can deeply alter the way people think about certain animals or their environment (Martin 1978; Pálsson 1994; Klemettinen 2002; Ylimaunu 2002). And vice versa, environmental change can further alter the way people perceive animals and nature (Martin 1978:144-149). Understanding the changes in subsistence activities as tied to both economy and worldview enriches our conceptions of human-environmental relationships in past societies (Hastorf & Johannessen 1996:75).

2.3. Towards a zooarchaeology of people

Zooarchaeology has grown to be an integral part of archaeological research during the latter half of the 20th century (Reitz & Wing 1999:15-31). The historical development of zooarchaeology is inextricably connected to the rise of processual or new archaeology, in which archaeology was seen as an objective hard science with rigorous testing of hypothesis, and scientific methodology was much emphasised (Reitz & Wing 1999:23-27; Bartosiewicz 2003:24; Hodder & Hutson 2003). The development of zooarchaeological methodology, in particular, gained momentum during this period (Reitz & Wing 1999:23-27; Bartosiewicz 2003:24). A common approach to archaeological animal bone finds has been the study of subsistence patterns and economy on a purely functional level and the emphasis has often been in the detailed reporting of species diversity, age and sex composition of animal populations and taphonomy, with little concern on further understanding the human societies creating the archaeological animal bone assemblages (Whittle 2003:82; Armstrong Oma 2007:44-46).

Faunal remains from historic sites have been analysed in growing numbers from the 1960's (Jolley 1983; Landon 2005). In Finland, however, historic site zooarchaeology commenced in the 1980's (Vuorisalo & Virtanen 1988), while the majority of the studies on the subject have been made during the recent decade (e.g. Tourunen 2003; 2008; Lahti 2006). The analysis of animal bone finds from historical sites has developed in tandem with prehistoric zooarchaeology, with some exceptions worth noticing. Although the same methodology is employed in prehistoric and historic site zooarchaeology, and the broader patterns of the development of these two subdisciplines have been similar (Landon 2005), the existence of written sources and theoretical foci of historical archaeology have somewhat influenced the choice of research questions and themes in historic site zooarchaeology. Especially socio-economic and ethnic variation in foodways, the market systems of animal-derived products and craft specialisation have received abundant attention in historic site zooarchaeology (see Deagan 1996; Landon 2005 for overviews).

There are major problems in the economical focus of traditional zooarchaeology and the rational model of explaining human behaviour which are prevalent in much zooarchaeological studies. These problems are by no means unique to zooarchaeological research, but are related to the general criticism aimed towards processual archaeology (see e.g. Shanks & Tilley 1987; 1994) and the tendency to interpret the apparently familiar post-medieval past in terms of present Western values and perceptions while, in fact, people in the past may have had completely different worldview (Tarlow & West 1999). I have argued elsewhere (Puputti 2009a) that the economic and functional approaches to zooarchaeological remains in fact represent our present-day worldview, in which animals and nature are seen as passive, exploitable economic resources, rather than any arguable interpretation of how people engaged themselves with the world in the past.

Brück (1999) has strongly criticized the use of western scientific rationale in archaeological interpretation. Her critique is aimed especially at the archaeological study of rituals, but has also a wider relevance in archaeology. She argues that people in the past acted rationally according to their own view of the world, and thus, we cannot evaluate the rationality of their actions in regard to our own view of the world. If people in a given culture in the past used hunting magic, they did so because according to their knowledge and view of the world it actually worked and they thought, for instance, that the guardian spirit of animals would give them a bag if they followed a certain ritual practice. In this kind of worldview, ritual actions cannot clearly be separated from practical or secular actions, and people may take actions that seem irrational to us but are perfectly appropriate according to their view of the world.

If we then consider the material remains of hunting in such a culture from the point of view of western rationalism, we risk arriving at a wrong conclusion, or, in the least, at a biased and reduced picture of past human-animal interaction.

Post-processual criticism and recent emphasis on social archaeology have called forth more socially and culturally oriented approaches in zooarchaeological studies. It has been acknowledged that cultural factors deeply affect the choice of food, the discard patterns of animal remains and the relationship between people and different animal species, and that simple functional and economical interpretations of archaeological animal bone finds fail to take into account the complexity of attitudes to food and animals (e.g. Jones 1998; Marciniak 1999; Fiore & Zangrando 2006; Mannermaa 2008a). Such zooarchaeological studies often rely heavily on anthropological studies of human-animal relationships and animal symbolism (Fiore & Zangrando 2006), historical data and iconographic and contextual analysis of archaeological finds (Jones 1998; Morris 2005; Pluskowski 2005).

The theoretical viewpoints presented in Chapter 2.1. have profound effects on the interpretation of zooarchaeological material. Many recent zooarchaeological studies (e.g. Jones 1998; Morris 2005; Pluskowski 2005; Fiore & Zangrando 2006) combine anthropological studies of human-animal relationships and animal symbolism, historical data and iconographic and contextual analysis of archaeological finds into a zooarchaeological interpretation. People in the past may have had different kinds of relationships with different groups of animals and a detailed zooarchaeological and contextual analysis may very well reveal some of these ancient cultural boundaries (Jones 1998; Serjeantson 2000; Soderberg 2004). Moreover, as the relationships between human and animals are formed via mutual interaction (Armstrong Oma 2007:62-76), the zooarchaeological evidence of day-to-day practices concerning animals is fundamentally linked to attitudes to animals.

It is perhaps self-evident that this does not mean the empirical evidence of species diversity and the physical characteristics of animals should be rejected or omitted. Rather, it is suggested that careful examination of the primary data is the fundamental basis of the cultural and social interpretation of the animal bone finds (Marciniak 2005:6). It has thus been suggested that archaeological animal bone finds should be considered, in tandem with other archaeological find groups, as evidence of living animals, human interaction with animals, the values and perceptions attached to animals and the social and spatial milieus in which animals lived together with people in the past (Marciniak 2005:1-7; Armstrong Oma 2007:19-20).

In this study, I will mainly draw upon the idea of human-animal relationships as constant mutual interaction between people and animals. Interaction with different animals was a part of the daily life of people in early modern Tornio.

It affected and was an inextricable part of the human-environmental relationship and the worldview of the people. I take the position that the relationships between humans and animals in early modern Tornio were created and recreated in everyday interaction with animals, and that the archaeological record can be used to trace this interaction. I interpret the archaeological animal bone finds as material evidence of past human-animal interaction in inextricable practical and belief-related levels.

3. Zooarchaeological methods

3.1. Description of the individuals

Taxonomic identification was conducted with the aid of the skeleton collection of the Zoological museum at the University of Oulu, archaeological reference collection of the Laboratory of Archaeology at the University of Oulu and published identification criteria, mostly Barone (1999) and Boessneck (1969). Specimens were identified to element, side and closest possible taxon. In addition to morphological identification, skeletal measurements were used in distinguishing two reindeer subspecies, wild forest reindeer (*Rangifer tarandus fennicus*) and semi-domesticated reindeer (*R. t. tarandus*) (Puputti & Niskanen 2009). The postcranial morphologies of these two subspecies are virtually identical and individual skeletal measurements overlap extensively. However, combinations of skeletal measurements, describing the relative size and shape of the bones, could be used in identifying these subspecies. Sets of measurements of the limb bones of modern wild forest reindeer and semi-domesticated were used to create logistic regression equations and discriminant functions that were applied to identification of the archaeological reindeer bone finds (Puputti & Niskanen 2009).

Measurements were taken of the bones of domestic animals and reindeer in order to estimate the size of the animals and temporal changes that possibly took place in their size, as well as to assess the sex of the individuals. All the measurements were taken of fused bones and according to von den Driesch (1976), unless stated otherwise. A digital Vernier calibre was used in conducting the measurements on bone ends, and a large, custom-made spreading calibre in measuring the greatest lengths of long bones. The measurements were taken to the nearest 0.01 millimetre.

The sizes of the animals can be compared using either the original measurements, or converting these measurements to comprehensible size estimates, such as withers height or body mass (Reitz & Wing 1999:172-177; Albarella 2001:55). As the conversions of original measurements always contain multiple source or error, such as individual and environmentally induced variation between the animals and biases in the material used to construct the conversion coefficients (Scott 1990; Bartosiewicz 1995:45), it is perhaps preferable to use the original measurements. Analysis of greatest lengths of bones and width and depth measurements can reveal patterns in the size and shape of the animals (Albarella 2001). However, estimates of withers height or body mass can be used to illustrate the magnitude and significance of change in body size. For instance, does a change of two millimetres in the mean humerus distal breadth convert to an observable change in body mass? Thus, both original measurements and body size estimations are presented and analysed in this study.

Epiphyseal fusion and dental wear and eruption were used in age assessment. The ageing data derives mainly from epiphyseal fusion, as the sample sizes were the largest. However, epiphyseal fusion is dependent on several factors such as nutrition, sex and population and thus, the estimated ages can deviate from the true biological ages of the individuals (Silver 1969; Bull & Payne 1982; Grigson 1982; Davis 2000). Dental eruption, on one hand, is probably less variable between populations and sexes, but on the other hand, is a useful ageing method only until the animals have obtained their final permanent dentition (Bull & Payne 1982; Bullock & Rackham 1982). Dental wear can complement data obtained from dental eruption. I employ the method of Grant (1982), in which the wear stages of mandibular molars are used to construct an estimation of mandibular wear stage. This method, however, is highly subjected to variation in the quality of the fodder (Sten 2004) and the length of the wear stages may differ (O'Connor 2000:88). All three methods are used, their results are compared and possible sources of error are discussed although it is acknowledged that it is impossible to precisely relate the age estimation to the true biological ages of the animals.

Sexual indicators and sexual dimorphism varies according to species, and different methods were thus used for different species. In this study, I primarily used pelvic morphology and metacarpal measurements in sex assessment of cattle and sheep or goat. The morphology of the pubic bone and the fossa muscularis of the musculus rectus femoris were used in sex assessment (e.g. Grigson 1982; Vretemark 1997:43; Greenfield 2006). In addition, the minimum thickness of the medial wall of the acetabulum of cattle was measured, a feature which has shown notable sexual dimorphism (Vretemark 1997:43-44; Greenfield 2006). All these features show considerable variation between individuals but are nevertheless reliable in sex assessment (Greenfield 2006). Canine morphology was used in sex determination of the pig (Mayer & Brisbin 1988).

Skeletal measurements of metacarpals were also used in sex assessment of cattle and sheep. The bones of the forelimb are often good sex predictors, as bigger proportion of the body weight, as well as the weight of horns or antlers, is deposited on the forelimb, although the degree of sexual dimorphism varies (Higham 1969; Grigson 1982; Klein & Cruz-Uribe 1984). Metacarpals are often used in archaeological sex assessment because of their durability: archaeological assemblages often contain complete metacarpals (Grigson 1982). Moreover, studies on modern cattle have confirmed the sexual dimorphism of cattle metacarpals (Grigson 1982; Berteaux & Guintard 1995; Svensson et al. 2008). Determination of sex from sheep metacarpals is, although an applicable method, more prone to error (Davis 2000; Tourunen 2008).

Human modification of bones, burning and animal tooth marks on bones were recorded in order to study depositional processes and animal carcass utilisation. Human modification was described as 1) cut marks: elongate, relatively narrow linear striations caused by removal of soft tissues attached to bones (Fisher 1995:12) 2) chop marks: broad, relatively short, linear depression generally with a V-shaped cross section, caused by separation of articulated bones (Fisher 1995:19), or 3) freshly broken bone: bone fragments with helical fracture lines, smooth fracture surface and sharp fracture angle to the cortical bone caused by bone breakage for marrow extraction (Villa & Mahieu 1991; Outram 1999; Outram 2001). The frequencies of each type of fracture were analysed. Only cattle and reindeer bone breakage was analysed as the bones of other species carried very little marks of human modification, perhaps because of the different body compositions of smaller species and cultural practices relating to the handling of the carcasses of these species.

3.2. Quantitative methods

Quantification is a central issue in zooarchaeology, because several factors such as differential preservation, bone fracturing, identifiability and sample size can considerably affect frequencies of species and body parts encountered in an archaeological animal bone assemblage. Numerous quantification methods have been developed to tackle these problems, and although none of the methods is flawless, the critical use of several methods is likely to yield a comprehensible picture of the species and body parts present in the animal bone assemblage.

NISP (number of identified specimens) is the fundamental and most often used quantification method. I follow here Grayson's (1984:16) definition of a specimen as "a bone or tooth, or fragment thereof" and the number of identified specimen includes all specimens identified to a taxon (Grayson 1984:16; Lyman 1994:100). NISP is prone to several sources of error, such as differential fragmentation of skeletons and problems associated with identifiability of certain species or skeletal elements (Grayson 1984:17-26; Reitz & Wing 1999:192-193). However, NISP is the original and fundamental means of quantification and is mainly used in my study when issues arising from problems associated with this method are not involved.

MNI (minimum number of individuals) is also often used in quantifying the representation of species. It is the smallest number of individuals needed to account for all the bone fragments of a species recovered from a given site (e.g. Lyman 1994:100). In this study, the age and sex of the specimens were taken into account when estimating the MNI, but the size of the specimen was not, as individuals can be asymmetric (Lyman 2006). MNI tends to overestimate the proportion or rarer species (Uerpmann 1973), which is problematic especially in this kind of an assemblage which includes several rare species. Especially in the early seventeenth century assemblage there are several game animal species, which are quite rare, and the use of MNI in quantifying this kind of assemblage would lead to severe overestimation of the share of game animal bones. Moreover, assemblage partition has a highly significant effect on MNI's (Grayson 1984:29-49), which causes problems when analysing urban archaeological bone assemblages that consist of numerous contexts and where the body parts of a single animal can be distributed across the town because of meat cut trade (Armitage 1982:95).

One of the central issues in this study is the species diversity, especially the diversity of hunted species. What makes comparing species diversities between excavation areas and periods difficult is that sample size greatly affects the proportions and numbers of species (Grayson 1984:116-167; Cannon 2001). The relationship of sample size and relative abundances is explored with basic chi-square test and Cochran – Armitage test for trend which also takes into account a trend in relative abundances (Zar 1996:562-565; Cannon 2001:193-194). The impact of sample size on species richness is analysed by using correlation and the log series α diversity index, which is perhaps less affected by sample size than many other diversity indices developed in the field of ecology, as well as easily converted to an understandable estimate of species richness (Krebs 1999:423; Magurran 2004:84-86).

The frequencies of meaty and non-meaty body parts of domestic taxa are compared. There are several possibilities in which the vertebrate skeleton can be divided to meaty (kitchen waste) and non-meaty parts (slaughtering waste), and in the end, all these divisions are dependent on the anatomy of the species and perhaps more importantly, cultural practices (Uerpmann 1973; Bartosievicz 1995:38-40). Here, I employ the division of During (1986), in which the cranium, caudal vertebrae, metapodials, carpals, tarsals and phalanges are classified as non-meaty body parts and vertebrae, scapula, pelvis and the long limb bones as meaty body parts, while acknowledging that also non-meaty body parts may have been valued. However, as the specialisation of meat producing and distribution often causes an overrepresentation of meaty body parts in some parts of the town and concentration of slaughtering waste in others (Maltby 1979:38-40; Landon 1996:8; Bowen 1998:138; O'Connor 2000:165-166), this rough model is used to analyse the presence and development of meat cut trade in Tornio.

Body part frequencies were examined in more detail as NISP and minimum numbers of elements (MNE) of each skeletal element. The MNE is, put simply, a minimum number of individuals calculated for each skeletal element (Reitz & Wing 1999:215-216), or, the minimum number of skeletal elements needed to account for the observed specimens (Lyman 1994:102). Lyman (1994:102) defines MNE as the "minimum number of a particular skeletal element or portion of a taxon". The choice of the skeletal element of portion is up to the researcher and depends on the research questions as well as the anatomy of the animal species. Quantification of body part frequencies using these methods also suffers from the same basic problems as quantification of species. As the MNI, the MNE is an analytical tool rather than an observational unit (Lyman 1994:102), i.e. the MNE does not mean that the elements are actually derived from the same individuals. Thus, I combined the skeletal elements from different excavation areas when counting the MNE. This was also necessary as the numbers of bones from each excavation area were not large enough for meaningful comparisons of frequencies of individual skeletal elements. In addition to MNE, also modified anatomical units (MAU) can be calculated for each element. This involves the dividing the MNE with the number of the given element in a complete skeleton, which enables the researcher to compare the frequencies of such elements as vertebrae and femurs, of which there are different numbers in the skeleton (Reitz & Wing 1999:215-216).

4. Archaeological material

4.1. Archaeological sites

Several archaeological excavations (Fig. 2) have taken place in Tornio from the 1960's on, most of them small-scale and small-budget rescue excavations (Ylimaunu 2007:17-18). Excavations have mainly been conducted in the central and southern parts of the town area, and mainly along the Keskikatu Street (Ylimaunu 2007:17). The animal bone material used in this study originates from the Keskikatu, Aspio and Viippola, Aho and Purra, Westring, and Välikatu excavations.

The Keskikatu excavations consisted of two modern-day plots that roughly correspond to seven plots of the early modern age. In all, ca. 1900 m² were excavated, but only a fraction of the area was properly studied (Herva 2003; Ylimaunu 2007:18). All the descriptions and dates of features encountered in the Keskikatu excavations are from the reports by Herva (2003) and Nurmi (2005a), and detailed maps of excavation areas can be found in Herva (2003). The excavation was conducted using the stratigraphic method. Sieving was not employed, due to a restricted budget and time, but the cultural layers were mainly removed using trowels. Stratigraphy, pottery, clay pipes and coin finds were utilised in dating the stratigraphic units (Nurmi 2005a).

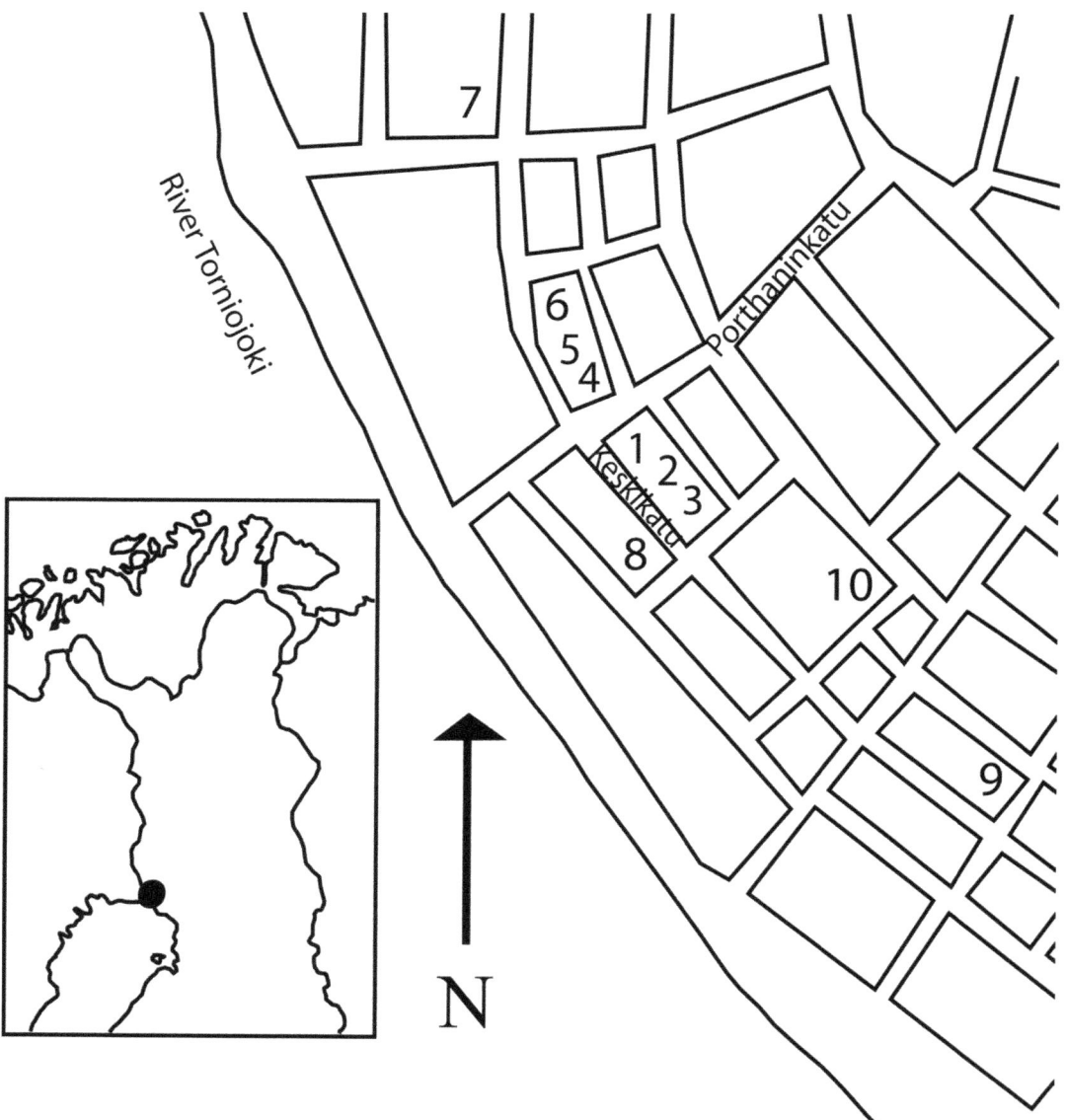

Figure 2. The location of Tornio and the excavation areas situated on the map of present-day Tornio. 1) Keskikatu excavation area 1, 2) Keskikatu excavation area 2, 3) Keskikatu excavation area 3, 4) Keskikatu excavation area 5, 5) Keskikatu excavation area 6, 6) Keskikatu excavation area 8, 7) Purra & Aho 8) Aspio & Viippola 9) Westring, 10) Välikatu. (Drawing A. Puputti)

In excavation area 1, the remains of two buildings, yard deposits and pit fillings were studied. The building in the centre of the area was interpreted as a residential building dating to the first half of the seventeenth century, while the function of the building in the northern part of the area is unclear. Building remains with unidentified function were found in area 2. In addition to building remains, a timber-covered yard, three pits and a timber-constructed cellar were found in this area. In area 3, the remains of a building, layers associated with a ditch and two pits were studied. The building on the southern side on the ditch was interpreted as a residential building that dates to the beginning of the seventeenth century. A timber-covered ditch which ran in the eastern corner of the excavation area in NW-SE direction is probably a sewer. In area 4, the remains of a lightly built building of unidentified function and a pit were excavated.

In area 5, the remains of a residential building with two rooms, a roofed pathway, two small cellar pits and a fireplace were studied. The building had probably two phases of use, which dated to 1620-1650 and 1660-1680, respectively. The remains of a lightly built wooden building which dates to the latter half of the seventeenth century, and a large yard deposit were also studied. In area 6, the remains of a residential building which dates to the early seventeenth century were excavated. Other remains underneath it were also documented. Their relationship with the youngest building phase is unclear, but it is probable that excavation area 6 has had two or three building stages. Remains of a building were excavated in area 8, but its function is unclear.

In the Aspio & Viippola excavation, the remains of seventeenth-century buildings, one of which was possibly a small animal shelter, were studied (Ylimaunu 1996; 2007:40-41). The overall excavation area was very small, only ca. 11 m² (Ylimaunu 2007:18). The animal bones from the Aho & Purra excavations were associated with the remains of a seventeenth-century smithy and a sewer, and the excavation area was ca. 30 m² (Ylimaunu 2001; 2007:18). In the Westring excavations, a large area was unearthed, although only ca. 30 of the total 100 m² were properly studied (Ylimaunu 2000). Remains of a building, possibly a smithy and a granary, and a sewer, dating to the eighteenth century, were found from this excavation (Ylimaunu 2000; 2007:18). All three locations were excavated as grid excavation (Ylimaunu 1996; 2000; 2001). In the Välikatu excavation, the remains of an eighteenth-century building were discovered. The function of this building is unknown (Nurmi 2005b). This excavation was done employing the stratigraphic method, and no sieving was done (Nurmi 2005b).

4.2. Overview of the bone material

The animal bone material in this study consists of ca. 21 000 bone fragments weighing ca. 200 kg. Most bone finds derive from the extensive Keskikatu excavations in 2002. Other excavations were smaller in scale and thus yielded less faunal remains. Most of the archaeological material dates to the seventeenth century. The spatial and temporal distribution of the animal bone finds are presented in Figure 3.

The lack of sieving is known to affect the species diversity of animal bone assemblages, and especially bones of small species, as well as minute part of skeletons of larger species, tend to be under-represented in un-sieved animal bone assemblages (e.g. Jones 1982; Shaffer 1992; O'Connor 2004:96-112). In the animal bone assemblage from Tornio, this is likely to affect especially the representation of small and fragile fish bones and bones of small birds and mammals. However, as the bones of relatively small species such as arctic hares and black grouse were abundant in the assemblages, it is possible that the majority of large and medium-sized game animal species are quite well represented in the trowelled archaeological material.

In general, the bones were in fairly good condition with little or no surface erosion. Very few complete bones of large mammals were found, probably due to breakage of bones

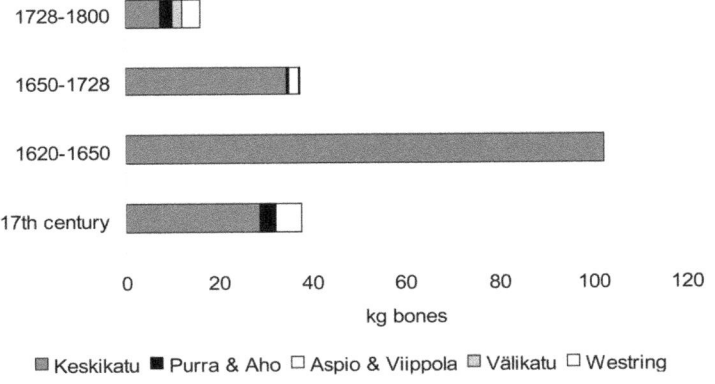

Figure 3. The spatial and temporal distribution of the animal bone material according to weight (kg).

during slaughter and food preparation. This issue is discussed in more detail in Chapter 5. Bones of medium-sized and small mammals and birds were often found whole. Burned bones were not common among the archaeological finds and they comprised only ca. 5 % of the total NISP. Gnawing damage by rodents and carnivores was minimal, which implies either lack of scavengers in the town or quick burial of waste in all context types (Puputti in press a). The preservation conditions in different context types were compared by analysing surface erosion, proportion of burned bone, mean fragment size and the representation of species of different sizes, and it seems there were no considerable differences in bone preservation between different context types (Puputti in press a).

4.3 Disposal of animal remains

It is probable that people had different attitudes towards different animal species and body parts of animals, and this is likely to be reflected in the treatment of their remains (Price 1985; Jones 1998). In the animal bone material from the Keskikatu excavation the clearest waste-disposal patterns emerged in the treatment of large-sized waste, bones of companion animals and treatment of bear claws (Puputti in press a).

In all, the animal bone material was quite homogeneous; no considerable differences in preservation conditions were observed and the remains of all species and all body parts were present in different context types (Puputti in press a). This may be due to the long formation periods and unknown formation processes of some of the archaeological contexts, and some degree of secondary deposition and disturbance of archaeological contexts. However, especially the pit and ditch fillings differed from contexts associated with buildings. They contained more parts regarded as slaughtering waste, and larger fragments of bone. Thus, it seems that pit structures and the ditch have been used as waste dumps. The finds from yard deposits and cellar fillings were more difficult to interpret, and complex formation processes including some degree of secondary deposition and long formation periods seem to have been prevalent. (Puputti in press a)

Companion animals, such as dogs, cats and horses seem to have received different treatment than meat-producing species. Their bones were rare in the archaeological assemblage (App. 1-3), although historical documents clearly indicate that for instance horses were present in the town (Mäntylä 1971:52, 120). However, a few disarticulated parts of skeletons of these species were found among the animal bone material (Puputti in press a). In the case of dogs and cats, individual bone finds probably imply that the bones in question are secondarily deposited bones of buried animals or that the deposits were disturbed (cf. Stallibrass 2000), since there are no signs of butchery or skinning in the bones. It is possible that cat and dog remains were intentionally buried in association with buildings, or that for example, a cat has died under a building. Another possibility is that sometimes their remains were discarded among with other refuse.

It is likely that horses were generally buried as whole carcasses (e.g. Jarva *et al.* 2001). Horse meat was traditionally not consumed in Finland, and in fact, killing a horse was generally such a taboo that the person finishing off horses was even not allowed to dine with other people and the dishes used by him were sometimes burned (Vuorela 1975:197). The horse bones in the Keskikatu assemblage were found in seventeenth century yard and underfloor deposits. They did not carry marks of skinning or disarticulation, but they have probably been used in preparation of soap or glue. All the horse bone finds were parts of the limb extremities, and considerably more eroded than other bone finds in the same context, which may be a sign of cooking the bones (Puputti in press a). Inferior parts of carcasses of animals have traditionally been cooked for soap or glue procuring (Talve 1990:101).

The distribution of bones of wild animal was not dramatically different from that of bones of domestic animal bones (Fig. 4). Most of the bone finds ca. 50 % of the bones of domestic animals and ca.70 % of the bones of wild animals) were discovered from contexts associated with buildings, and smaller amounts from pit, ditch and cellar earth fills and yard deposits. Bones of wild animals were more common than those of domestic animals in contexts associated with buildings, and rarer in pit and ditch fillings. However, it is difficult to determine whether the observed disposal pattern is because of distinct waste disposal practices for the remains of domestic and wild species, or simply because the bones of most domestic animals are larger and thus more noxious as waste than those of small wild species. To address this problem, the proportions of bones of small species (wild gallinaceous birds, waterfowl, domestic chicken and small wild mammals) in different context types and the proportions of bones of middle-sized animals (ovicaprids, pigs and seals) were analysed (Fig. 5), and it does seem that wild and domestic species of the same size category were treated more or less the same way. Approximately 70 to 80 % of bones of small species were deposited to building-related contexts. Of the bones of middle-sized animals, ca. 60 % were discovered from contexts associated with buildings.

The deposition of a set of bear claws under a foundation timber of a residential building in Keskikatu excavation Area 1 infers to ritual practices. The bear has been perceived as a powerful animal in Finnish folklore, while at the same time, it was considered harmful to cattle and bounties were offered of killed bears (Klemettinen 2002). Foundation deposits containing animal bones are, although not abundant, frequently reported from archaeological sites in Finland and Scandinavia (Hukantaival 2007; Falk 2008). Especially limb bones of animals were often deposited under the foundations of buildings, and these were most likely believed to protect the building from ill will (Hukantaival 2007). Other such 'special' deposits including animal bones were not discovered from the excavations in Tornio. It is likely, however, that beliefs and ideas have been attached to other animals and body parts of animals as well, while these may have not left traces in the urban animal bone material.

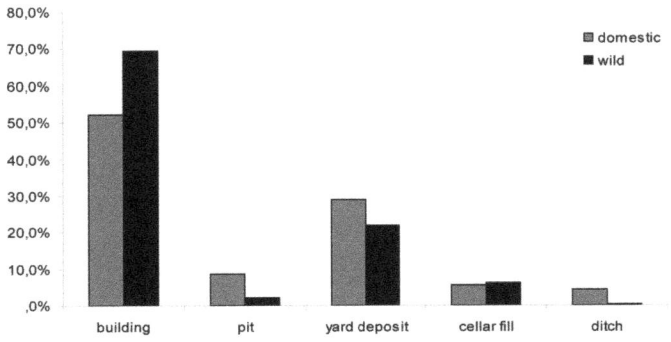

Figure 4. The distribution of bones of wild and domestic animals in different context types in the Keskikatu excavations.

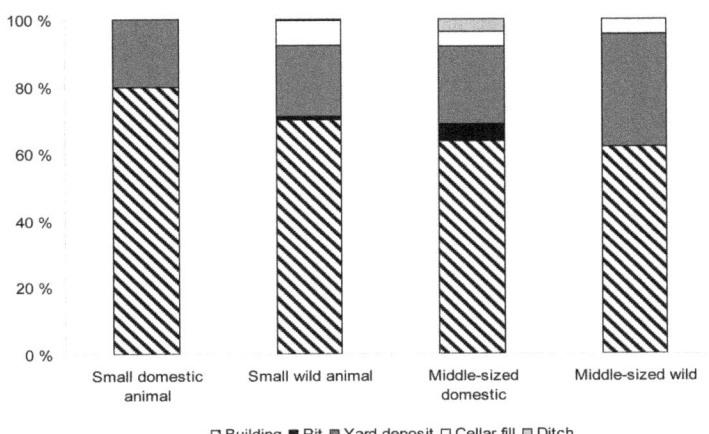

Figure 5. The distribution of bones of different-sized animals in different context types in the Keskikatu excavations.

5. Analysis of the bone material

5.1. Patterns in species diversity

The relative abundances of bone fragments identified to species or genus are presented in Table 1, whereas the abundances of all taxa, divided by period and excavation area are presented in Appendices 1-3. The animal bone assemblage from early modern Tornio is dominated by cattle (*Bos taurus*): cattle bones constitute ca. 40 to 70 % of the specimens identified to species and genus, depending on the location and period. Also sheep (*Ovis aries*) or goat (*Capra hircus*) and pig (*Sus scrofa domesticus*) bones are relatively common, the shares of their bones being 13-19 and 4-9 %, respectively. Distinguishing sheep and goat is extremely difficult from most skeletal elements, while some elements, such as complete metapodials are more easily identified (Boessneck 1969). In the assemblage from Tornio, only five specimens were identified as goats. The goat bones derived from a seventeenth century deposit in excavation area 8 and a late seventeenth century deposit in excavation area 3. Four of these were metapodials and one was the back of the skull, both elements that are fairly easily identified to species (Boessneck 1969; Barone 1999). As discussed in chapter 4.3., the bones of companion animals were very rare.

Reindeer (*Rangifer tarandus*) bones constitute 1-6 % of the assemblage and the exact proportions of wild forest reindeer (*R.t.fennicus*) and semi-domesticated reindeer (*R.t.tarandus*) are unknown. However, both these subspecies may have been present in early modern Tornio, as reindeer husbandry was practiced in the area and several people in the town owned reindeer, and on the other hand, wild forest reindeer were hunted (Mäntylä 1971:109; Virrankoski 1973:243-246). With the aid of discriminant analysis and logistic regression of postcranial skeletal measurements, a number of reindeer bone finds could be assessed to subspecies within the 95 % confidence level (Puputti & Niskanen 2009). Of these 12 assessments, nine were semi-domesticated reindeer and three wild forest reindeer.

Wild mammals and birds were hunted in abundance, especially during the seventeenth century, their share being ca. 30 % in the early seventeenth century. Arctic hares (*Lepus timidus*) and seals (*Phoca* sp.) were the most common wild mammal species. In most cases, the seal bones could not be identified to species. Six ringed seal (*Phoca hispida botnica*) bones and two grey seal (*Halichoerus grypus*) bones were identified among the archaeological material, and it is likely that the other finds also belong to one of these species, as they inhabited the Bothnian Bay during the early modern period (Ylimaunu, J. 1996:182; Ukkonen 2002). Additionally, a few bones of red squirrel (*Sciurus vulgaris*), red fox (*Vulpes vulpes*) and brown bear (*Ursus arctos*) were also found. Large herbivores, such as elks (*Alces alces*) and wild forest reindeer, were apparently also hunted. In addition to the three wild forest reindeer bones, also one proximal humerus of elk was found. Large game may have been hunted more intensively than the animal bone assemblage indicates, but there bones may have been left to the kill site and only meat brought to the town.

Bones of wild gallinaceous birds were common in the archaeological assemblage, their bones making up 3-15 % of the bone fragments identified to species or genus. The most common species was capercaillie (*Tetrao urogallus*), followed by the black grouse (*T.tetrix*), willow grouse (*Lagopus lagopus*) and hazel grouse (*Bonasa bonasia*). Also waterfowl were hunted in abundance and their bones were especially numerous in excavation area 5 during the early seventeenth century. Different species of geese (*Anas* sp.), as well as whooper swans (*Cygnus cygnus*), mergansers (*Mergus* sp.) and other anatids (*Anas* sp.) were hunted.

Table 1. Relative abundances of all bone fragments identified to species or genus (% NISP).

	% 1620-1660	% 1650-1728	% 1720-1800
cattle	41	49	70
sheep or goat	19	18	13
pig	7	9	4
reindeer	2	1	6
other domestic animal	<1	<1	0
seals	3	2	1
wild mammal	7	7	3
gallinaceous birds	15	10	3
waterfowl	6	5	1
Total N	2879	739	252

Fish bones were found in moderate amounts. The interpretation of the fish bone finds, however, is hindered by the fact that sieving was not employed in the majority of the excavations, and thus, the fish bone finds are likely to be biased towards especially large species and skeletal elements. Most of the fish bones seem to belong to salmonids (*Salmonidae*) and also some cyprinid (*Cyprinidae*) bones were found, but fish bones are in large part excluded from this study as the fish bone assemblage is probably considerably biased.

The skeletal frequencies of wild animals, divided by period, are presented in Appendix 4 as NISP, MNE and MAU. Skeletal frequencies of wild mammals (excluding the elk and brown bear because of the minimum number of elements and minimum number of individuals of one), indicate that all body parts are represented. The frequencies of the extremities are not especially high in arctic hare, seal and red fox, an observation that indicates that these animals were brought to the town as whole carcasses, not as furs or skins. The skeletal frequencies of wild birds resemble a natural assemblage; the robust bones of shoulder and wing (especially the coracoid bone and humerus) are most abundant in most species, whereas the bones of the hindlimb, axial skeleton and cranium are less abundant (Tagliacozzo & Gala 2001; Mannermaa 2008b:48). Crania and extremities were represented especially in large species, and the fragile cranial bones and small bones of extremities of smaller birds have probably been lost in the excavation. It is difficult to make any explicit conclusion on the skeletal representation of each species, as most of the sample sizes are quite small. The data at hand indicates that whole carcasses of most wild animal species were brought to the town and this suggests that the townspeople were themselves involved in hunting. Of course, small animals such as hazel grouse and hares may have been traded to some degree as whole carcasses, but the representation of crania and extremities of large game birds such as geese and swans on one hand, and the lack of overrepresentation of lower limb bones of fur-bearing animals and seals on the other hand supports the idea that the urban archaeological animal bone finds are a product of the hunting activities of the townsfolk.

In all, hunting activities seem to concentrate on inland game animals, apart from fishing. Although waterfowl were hunted, especially during the seventeenth century, and the proportion of seal bones remains small but constant throughout the studied period, wild gallinaceous birds and arctic hares clearly outnumber aquatic species. This is rather puzzling, given the central role of fishing in the economy of Tornio (Mäntylä 1971:121). It seems that apart from fishing, aquatic resources did not play an important role in the subsistence and economy of the townsfolk, at least from the late seventeenth century on. Animals with a valuable fur, i.e. foxes, squirrels, wolverines etc., are extremely rare in the animal bone assemblage from Tornio, whereas arctic hares, less valued for their fur, are quite common. This emphasises the role of hunting as a subsidiary mode of subsistence and that valuable trade items such as furs, blubber and hides were probably acquired via trade relationship with the Sami and the farmers.

The diversities of domestic species varied between the studied plots. The proportion of cattle bones of all bones of domestic animals varied considerably. The shares of cattle bones varied between 53 and 63 % in the early seventeenth century, between 58 and 74 % in the late seventeenth century and between 74 and 79 % in the eighteenth century (Fig. 6).

Also, there are some differences in the frequencies of wild animal bones and numbers of hunted wild species within the town. In the early seventeenth century, the shares of wild animal bones of all the bones identified to species or genus vary from 9 to 38 %, from 7 to 37 % in the late

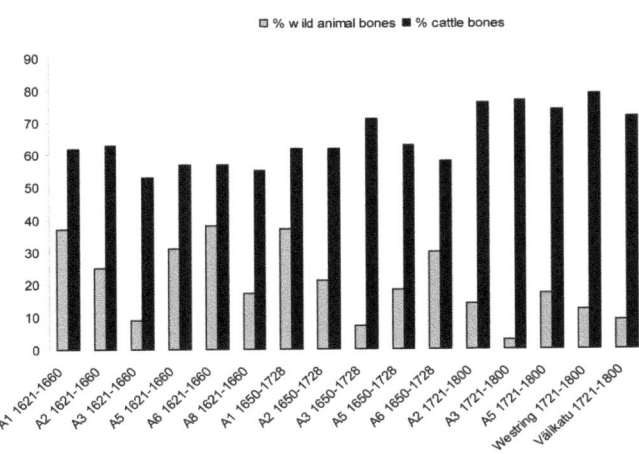

Figure 6. The proportions of bones of wild animals of all bones identified to species or genus (% NISP) and proportions of cattle bones of all bones of domestic animals (%NISP) in different excavation areas and periods.

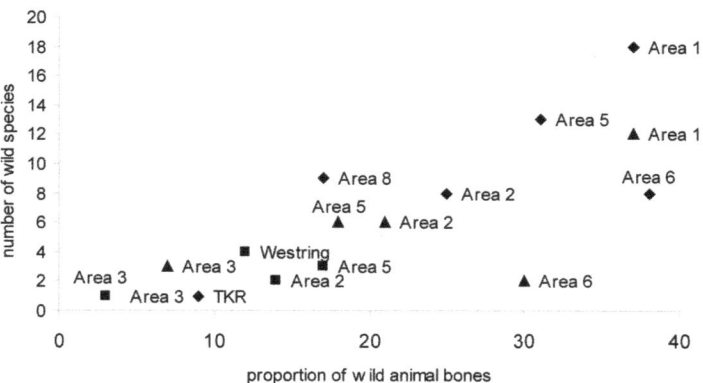

Figure 7. The proportions of wild animals of all the bone fragments identified to species or genus (% NISP) plotted against the number of wild species encountered in the assemblage. Key: □ *1620-1660* ▲ *1650-1728* ■ *1721-1800.*

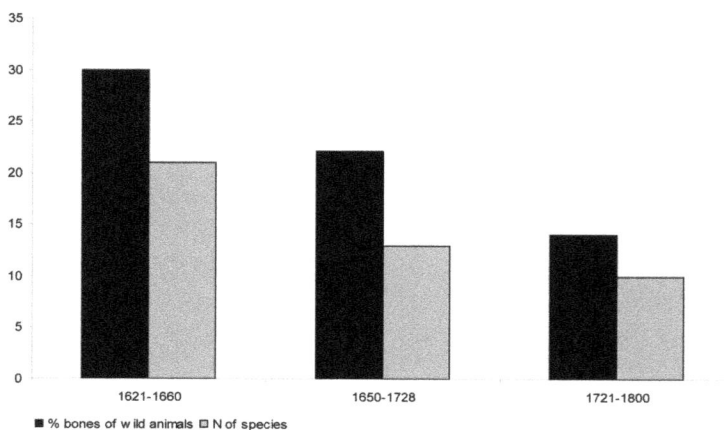

Figure 8. Proportions of bones of wild animals of all animal bones identified to species or genus in different periods, and the number of wild species in each period.

seventeenth century and from 3 to 17 % in the eighteenth century (App. 1-3). The numbers of wild species in the assemblage range from 1 to 18 in the early seventeenth century, from 2 to 12 in the late seventeenth century, and from 1 to 4 in the eighteenth century (App. 1-3). In general, in excavation areas with a high proportion of wild animal bones, also the diversity of the hunted species was larger (Fig. 7). Only Keskikatu excavation area 6 seems to be an exception, as this area yielded a high proportion of wild animal bones but only a modest number of species (Fig. 7). Area 6 excluded, the correlation between the proportion of wild animal bones and the number of wild species is very high (r = 0.92 p = 0.01). As the number of species encountered in a bone assemblage is highly dependent on sample size (Grayson 1984:116-167), the different sample sizes may affect this result somewhat. However, the correlation between sample size and species diversity was lower than the correlation between the proportion of wild animals and the number of species (r = 0.85 p = 0.01). Furthermore, the chi square test of the wild animal bone ratios in different excavation areas and periods gave highly significant results (χ^2 = 260.3 p = 0.000), indicating that the share of wild animal bones does indeed differ, even when different sample sizes are taken into account. Thus it seems that, in general, the people actively engaging in hunting may also have exploited a wider variety of species.

There are some changes in the species diversity between the early seventeenth century, the late seventeenth century and the eighteenth century (Puputti 2008). Especially the proportion of bones of wild animals decreases markedly and cattle comes to dominate the assemblage. It changes from the c. 30% in the beginning of the seventeenth century, to the 22% in the late seventeenth century and to the 14% in the eighteenth century (Fig. 8). Also, the diversity of the hunted species decreases dramatically: in the assemblage dating to the early seventeenth century the number of game animal species is 21, most of them waterfowl, whereas in the eighteenth century the number of species is only 10 (Fig. 8). In all, the subsistence base seems to narrow down during the eighteenth century (Puputti 2008).

It is widely acknowledged that sample size may considerably affect relative abundances and species richness (Grayson 1984:116-167; Cannon 2001). In short, larger assemblages tend to have more species up to a certain point where the number of species does not increase anymore although the sample size increases (this is called species-area relationship) (Magurran 2004; Lyman 2008:180). The proportions of species in the assemblage may also be biased in small assemblages (Cannon 2001). Hence, the different sample sizes in comparing excavation areas and periods present a possible source of error. I used the chi-square and Cochran-Armitage tests, the latter of which is associated with linear trend (Zar 1996:562-565; Cannon 2001; Lyman 2008:201, to examine the relationship between sample size and proportions of wild species in each time period. Both tests gave highly significant results, indicating that the proportion of bones of wild animals indeed decreases ($X^2 = 64.2$ p = 0.000, $X^2t = 4.2$ 0.05 >p<0.01, $X^2d = 60.0$ p<0.001). There are a number of diversity indices to choose from to study the relationship between species richness, sample size and sample heterogeneity, and all these indices are somewhat influenced by sample size (Grayson 1984:158-167; Magurran 2004). Here, I used the log series α diversity index, which is fairly straightforward, measures the effect of sample size on species richness and is easily converted to understandable estimates of species richness (Krebs 1999:426; Magurran 2004:30). The index showed that the diversity of wild species was indeed higher in the early seventeenth century than in later periods (see Puputti 2008 for details). Based on these measures, it seems reasonable to say that the trends detected in the archaeological material are probably not caused, at least not solely, by different sample sizes. Thus it seems that, in general, the people actively engaging in hunting may also have exploited a wider variety of species, and that the species diversity and share of bones of wild animals indeed decrease especially in the eighteenth century animal bone assemblage.

5.2. Animal husbandry practices

Cattle

Age profile of cattle

The age structure of the cattle (*Bos taurus*) bone finds was studied by examining epiphyseal fusion and combined data on tooth eruption and wear (Fig. 9). Mandibular wear stages (M.W.S) (Grant 1982) were used to narrow the range of possible ages for individuals for whom tooth eruption could give only indicative estimations, such as, for instance, over 24 months. The tooth eruption stages and mandibular wear stages of all the cattle mandibulae from early modern Tornio are presented in Appendix 5. According to the data extracted from cattle teeth, it seems that mainly older individuals were slaughtered. There were only four individuals under six months of age, and four under 15 months. The majority of the killed individuals are ca. one to three years old, but a considerable number was older, over three years. It is not clear, what does the M.W.S. of over 40 points mean in years in the material from early modern Tornio. However, according to Vretemark (1997:84), medieval Swedish cattle with mandibular wear stage of 39 to 45 points, estimated by the same method, were four to eight years old, and those with M.W.S. of over 45 points were over eight years old. In the assemblage from Tornio, too, there seems to be a considerable number of old cattle, although the exact biological age is impossible to estimate.

The killoff pattern of cattle was examined also by studying the epiphyseal fusion. The advantage of this method is that epiphyseal fusion continues up to the age of 4.5 – 5 years in modern cattle (Barone 1999:76) and thus, this method is informative also in older age groups than tooth eruption. Moreover, bones suitable for age assessment with this method were far more numerous than mandibulae and thus, the larger sample size allows statistically more reliable results. The disadvantage of epiphyseal fusion as aging method is its variability according to breed, nutrition and sex, among other things (see e.g. Davis 2000). The epiphyseal fusion dates used in this study are extracted from modern breeds (Silver 1969; Barone 1999). The results, divided by period are presented in Figure 10.

Both dental and epiphyseal data indicate that most cattle were slaughtered as adults in early modern Tornio. In the youngest age groups, only a few percents of animals have been killed, and some of these very young individuals may also be the prematurely died calves. The first slaughtering period seems to be in the age of 1.5 – 2.5 years, or a little later if dental eruption and epiphyseal fusion have been delayed in early modern cattle in comparison with modern breeds. Still, a considerable number of animals reached the age of over 5 years, according to numerous fused vertebrae in the assemblages.

This kind of killoff pattern relates to emphasis on milk production in cattle husbandry, although the number of subadults is also relatively high. And indeed, the examination of the presentation of sexes in the cattle bone assemblage revealed a more complex pattern of sex-specific age distributions and culling patterns for cattle (see next chapter). Historical sources and other zooarchaeological analyses constantly emphasise that cattle in early modern northern Finland were mainly milk-producers, only eaten when they reached an old age (Virrankoski 1973:239; Wilmi 2003:178; Puputti 2007), and the historical record from Tornio indicates a dominance of older cows over calves and heifers in the town (Mäntylä 1971:52, 121). There does not seem to be differences in cattle age profiles based on epiphyseal fusion between the periods, which indicates that raising cattle solely for meat did not become common in Tornio during this period, although in more

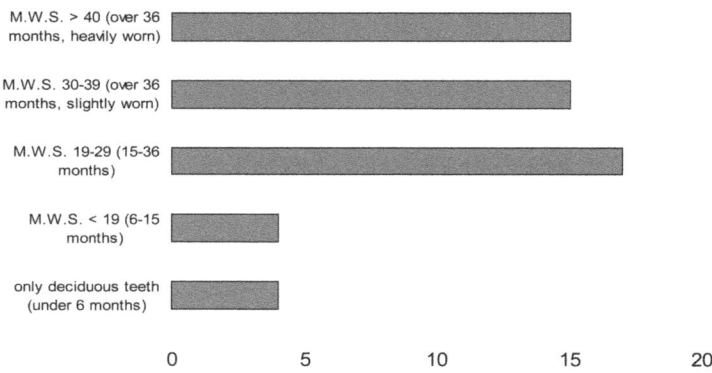

Figure 9. Combined data on tooth eruption and wear of cattle.

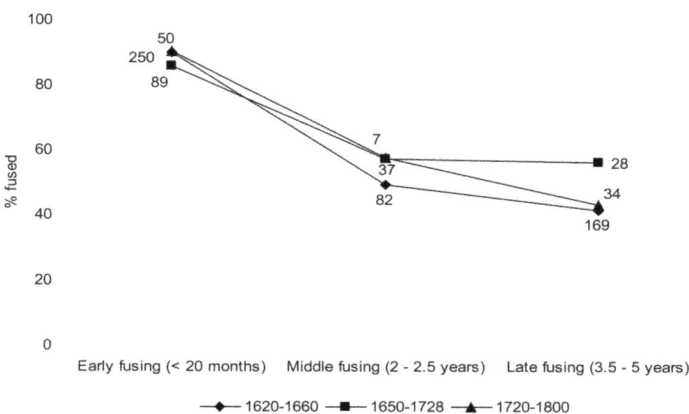

Figure 10. Proportions of fused epiphyses in each age group, divided by period. The early fusing elements are proximal scapula, acetabulum, proximal radius, distal humerus and proximal 1st and 2nd phalanges. The middle fusing elements are distal metapodials and distal tibia. Late fusing elements are femur, proximal calcaneum, proximal humerus, distal radius, proximal tibia, proximal ulna and vertebrae. Total numbers of anatomical elements are also presented.

southern towns of Finland and Sweden, there are indications of growing interest on meat production already during the medieval period (Vretemark 2003:88-89; Tourunen 2008:136).

Sexing cattle bones

Cattle bone finds were sexed with the aid of morphological and metrical methods in order to study cattle husbandry practices in early modern Tornio. The morphology of cattle pelvic fragments and the minimum thickness of the medial wall of the acetabulum were examined. Results are presented in Appendix 6. A total of 25 male and 54 female cattle pelvises were identified, and thus, pelvic morphology indicates the presence of both sexes in the assemblage, although male pelvic fragments constituted only a third of the material.

In addition to pelvic morphology, also sexual dimorphism of cattle metacarpals was studied. Cattle metacarpals are sexually dimorphic because the forelimb carries a bigger share of the body weight, which is considerably bigger in males, and they are often used in archaeological sex assessment because of their durability (Grigson 1982). Cow metacarpals tend to be small and slender, bull metacarpals large and heavy, and those of steers relatively long but slender (Higham 1969; Grigson 1982). In studies of modern cattle, the distal metacarpal alongside with metacarpal greatest length, have proven to be the most useful measurements for sex assessment (Grigson 1982; Berteaux & Guintard 1995; Svensson *et al.* in 2008). Here, I plotted the breadth of the distal end against the depth of the distal end (Fig. 11). The data on measurements on fused cattle metacarpals are presented in Appendix 7.

The scatter diagram of metacarpal measurements (Fig. 11) does not clearly indicate the presence of two or more groups. Only one specimen seems to be considerably bigger than the others, possibly male. The sizes of the cattle metacarpals from Tornio are very small in comparison with most other datasets of cattle measurements and fall in the range of females in medieval Swedish samples (Svensson et al. 2008) as well as in Central European bone assemblages (e.g. Bartosiewicz 1995:46). Taken the fact that cattle in early modern Tornio were very small (see next chapter) it could be possible that also males would fit in the female range of measurements from other parts of the world where cattle were larger. This problem was assessed by examining the slenderness indices of the metacarpals from Tornio. The relationship of joint size and bone length may be a good discriminator of sex because the lenght of the bone can exhibit considerable phenotypic plasticity, whereas joint sizes exhibit little phenotypic plasticity (Niskanen 2006). According to Berteaux & Guintard (1995:106), the metacarpal slenderness indices, calculated as 100*(mcBd/mcGL), in modern and archaeological cattle populations, are in most cases under 30 for females and over 30 for males. As the slenderness indices for cattle metacarpals from Tornio were in most cases considerably smaller than 30 (App. 7), it seems that majority of the metacarpals indeed derive from female animals.

Thus, there is a clear discrepancy between the pelvic and metacarpal data. As the age profiles indicate an emphasis on milk production in early modern Tornio, one would expect cows dominating the assemblage. The relatively large proportion of male cattle in the pelvic fragment assemblage, then, could be an indication of the use of draft

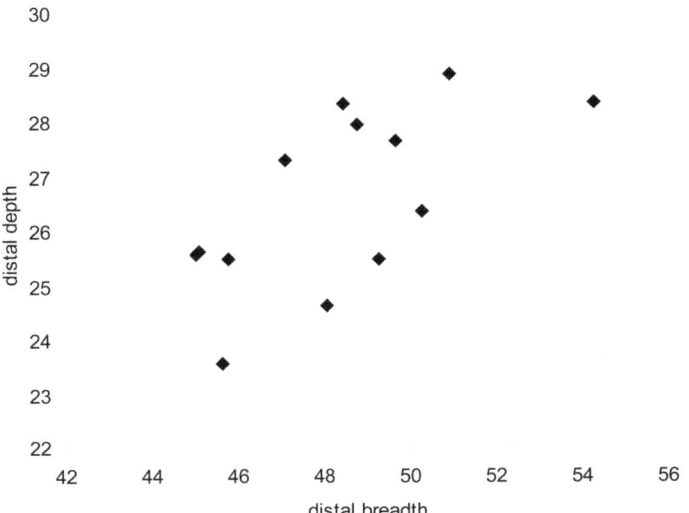

Figure 11. The scatterplot of the distal breadth and depth (mm) of cattle metacarpals.

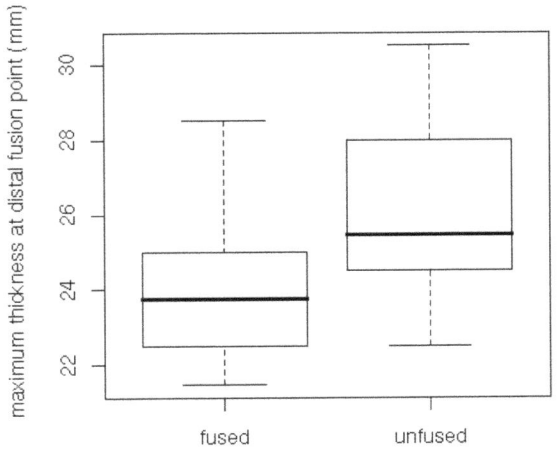

Figure 12. The boxplot of the maximum depth of fused (N= 14 ± 1.9) and unfused (N = 10 ± 2.6) cattle metacarpals at the distal fusion point (mm). T-test results are t = -2.4 p = .03.

cattle. However, historical and ethnographic data imply that draft oxen were probably not used in northern Finland (e.g. Virrankoski 1973:240). Additionally, one would expect also the metacarpal measurements to indicate the presence of three sexes, not one, if draft oxen were used. Thus, a more possible explanation for the phenomenon is that cows were used for dairying and slaughtered at an old age, whereas bulls were slaughtered for meat as subadults. The acetabulum of the pelvis fuses at an early age, at the age of seven to ten months in modern cattle breeds, whereas the distal metacarpal fuses at the age of approximately two years in modern cattle (Silver 1969:285-286; Barone 1999:76). This would explain why there are bulls in the fused pelvic fragment assemblage and predominantly cows in the fused metacarpal assemblage.

This question was analysed in detail by examining the dimensions of fused and unfused distal metacarpals. The maximum thickness of the distal fusion point was measured in fused and unfused cattle metacarpals (see Higham 1969 for measurement definition). This measurement was chosen, as it was possible to take from both fused and unfused samples, unlike, for instance, greatest length. Also, as majority of the metacarpals were craniocaudally split, width measurements were in many instances unobtainable.

As Figure 12 shows, the thickness of some of the unfused metacarpals was in fact bigger than those of mature, fused bones. Although the sample is relatively small and thus the results are not statistically highly significant, this could indicate that bulls were indeed slaughtered at a younger age than cows, which were used in dairying. As we examine the age profile extracted from epiphyseal fusion (Fig. 10), we see that the majority of the early-fusing epiphyses (fusing at the age of less than 20 months) are fused, while ca. 50 % of the middle-fusing elements (fusing at the age of 2-2.5 years) are unfused. Thus, it is possible that bulls were killed after the fusion of the early-fusing elements and before the fusion of metacarpals, which would translate to the age of 20-24 months in modern cattle. Thus, the cattle bone assemblage from Tornio seems to consist of adult dairy cows and approximately 7 to 24 months old bulls that were slaughtered for meat.

Size of cattle

Long bone greatest lengths are most often used in withers height estimations (von den Driesch & Boessneck 1974). In this study, I will use cattle metacarpals in withers height estimations, as only metapodial greatest lengths were measurable from the fragmented archaeological material. Several authors have created coefficients for converting metapodial greatest lengths to withers heights (see von den Driesch & Boessneck 1974 for a more detailed discussion). Here, I chose to use Matolcsi's (von den Driesch & Boessneck 1974) coefficients, created from a set of modern Hungarian cattle metapodials, because they have proved quite accurate in withers height estimations on a variety of cattle breeds (Berteaux & Guintard 1995:107). In addition, Fock's (von den Driesch & Boessneck 1974) coefficients were used in order to compare the size of North Finnish cattle with Swedish cattle from different periods, which were previously estimated with that method (Sten 1994).

Because of sexual dimorphism cows and bulls have separate conversion coefficients for withers height estimation. As discussed above, all but possibly one cattle metacarpal from Tornio seem to belong to female individuals. Thus, female conversion coefficients were used in all cases but this one individual, which was assessed as male. In the case of metatarsals, sex could not be assessed because the slenderness indices in metatarsals are much more overlapping than those of metacarpals (Berteaux & Guintard 1995:106) and thus, sex could not be assessed from metatarsal measurements in any certainty.

Cattle withers height estimations from metacarpals from early modern Tornio are presented in Table 2. The average cow withers heights, 102.2 and 101.3 cm, are quite small, and the Finnish traditional cattle race was known to be considerably smaller than its Central European counterparts (Vuorela 1975:197), although exact measurements are not available. The average withers height estimations for Swedish late medieval and early modern cows are somewhat higher than those for cows in Tornio, being ca. 110 cm (see Sten 1994), while those for cows in medieval Turku were ca. 104-106 cm (Tourunen 2008:91).

Body weight estimations have been attempted less frequently than withers height estimations for cattle bones in archaeological assemblages (e.g. Noddle 1971), although

Table 2. Withers height estimations of cattle based on metacarpal greatest lenght, estimated with the coefficients of Matolcsi and Fock (von den Driesch & Boessneck 1974).

Sex	Coefficient	N	Mean (cm)	Stand.dev.	Range (cm)
Male	6.33 (Matolcsi 1970)	1	115.2		
Female	6.05 (Matolcsi 1970)	11	102.2	5.1	93.2 – 108.6
Male	6.00 (Fock 1966)	1	113.7		
Female	6.25 (Fock 1966)	11	101.3	5.0	94.2 – 107.7

in palaeontology these are common (see e.g. Damuth & MacFadden 1990). In ungulates, the non-length dimensions of postcranial bones have proved to be the most accurate predictors of body mass (Scott 1990:301). Here, I applied the regression equations of Scott (1990) for bovids to cattle bone finds from Tornio. The data on original measurements of postcranial bones of cattle, ovicaprids, pig and reindeer are presented in Appendix 8. It seems that inter-taxon body mass predictions can be relatively accurate in as wide as a group as all ungulates, and highly accurate in the level of families (Scott 1990). I applied the equations to distal humerus, proximal radius and distal tibia, which were relatively abundant in the bone assemblage from Tornio, and which also had relatively low rates of estimation errors (Scott 1990). The results are presented in Table 3.

The resulting mean body weight of cattle in early modern Tornio, between 143 and 165 kg, is very small in comparison with, for instance, the live body weight of the traditional North Finnish cattle race in the beginning of the 20th century, which was ca. 300 kg (Grotenfelt 1916:16). Modern North Finncattle males and females weigh ca. 650 and 400 kg, respectively, with withers heights of 128 cm and 118 cm, respectively (www.tiho-hannover.de). The regression equations for bovids may underestimate the body mass of large bovids (Scott 1990:314), and thus, it is possible that the predicted body weights of cattle are slightly too small. The prediction error, however, should be within 20 % of the actual masses (Scott 1990:314). One must bear in mind, however, that Scott's regression equations are based on a sample of wild species (Scott 1990). Hence, their applicability to domestic animals can be questioned, as the body proportions of domestic animals may be expected to deviate from those of wild species, because they have been affected to selective breeding favouring, for instance, overall size or fat content of the muscle (Clutton-Brock 1987:22-25).

The sample sizes of individual skeletal elements were too small for statistically significant comparison between the periods. In order to increase the sample size, the log-ratio method was employed. In the log-ratio method, all measurements are converted into logarithms and they are subtracted from the log of a standard individual or sample (O'Connor 2003:180-181; Thomas 2005:78). This method permits to combine different measurement from the same species and thus the sample size increases (Thomas 2005:78). The choice of the standard specimen is a debated matter, others preferring a modern individual of known age and sex, and others relying on modern or archaeological population means (Albarella 2001:55). Here, the standard was extracted from archaeological finds from Oulu Pikisaari excavation, as no complete modern reference individuals were available. The log ratios of skeletal measurements of cattle, sheep or goat, pig and reindeer are presented in Appendix 9.

The comparison of cattle width and depth measurements with log ratio method did not show statistically significant differences in cattle size between the periods (Puputti 2008). When greatest lengths were included in the comparison, the examination revealed statistically significant differences between the periods. The size of cattle, especially in long bone lengths, seems to somewhat increase in the late seventeenth century and during the eighteenth century. The results are presented in Figure 13. This indicates a change in the body composition of the animals, perhaps larger withers heights and slenderer limbs.

The magnitude of the change in the body proportions of live animals, however, seems to have been quite insig-

Table 3. Body mass estimations of cattle based on skeletal measurements, according to regression equations of Scott (1990).

Measurement	N	Mean (kg)	Stand.dev.	Range (kg)
hum Bd	19	165.7	10.2	149.8 – 186.3
hum BT	17	152.5	11.3	138.6 – 176.9
tib Bd	13	143.3	6.7	135.7 - 157.0
rad Bp	19	162.5	11.3	145.8 – 187.6
all body mass estimations	68	157.2	13.1	135.7 - 187.6

Table 4. Comparison of withers height and body mass estimations (means, sample sizes and standard deviations) of cattle between periods.

	1620-1660	1650-1728	1721-1800
withers height (cm)	102.1 / N=4 / ±4.1	100.3 / N=3 / ±7.5	105.4 / N=4 / ±2.7
body mass (kg)	151.2 / N=31 / ±11.1	156.5 / N=9 / ±12.5	154.2 / N=5 / ±14.3

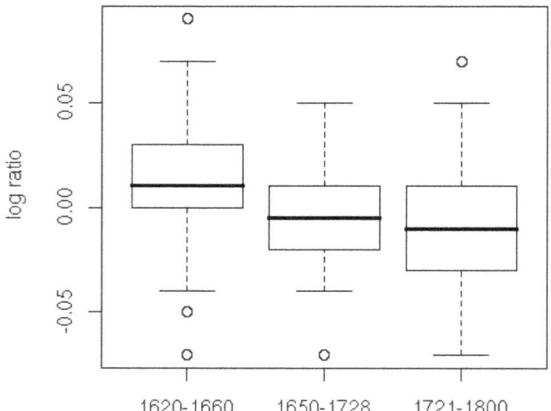

Figure 13. Comparison of log ratios of cattle skeletal measuremens between the periods 1620-1660 (N= 124 ± 0.03), 1650-1728 (N= 48 ± 0.02) and 1721-1800 (N= 43 ± 0.03). Two-way ANOVA results are F = 9.8 p = .000.

nificant. This is well illustrated in the withers height and body mass estimations divided by period (Table 4). The differences in estimated withers heights are only a couple of centimetres and the mean body masses fluctuate only a few kilograms. These estimations are of course merely approximations and the sample sizes are, again, very small. Cattle seem to be somewhat larger in the late seventeenth century and early eighteenth century, but the magnitude of the change does not imply any systematic improvement of cattle for meat production, or considerably better conditions and nutrition. Alternatively, the seeming increase in cattle size may in fact be caused by extremely small individuals in the sample from the early seventeenth century. Several serious years of crop failure and famine occurred during the seventeenth century (Virrankoski 1973:205-221), which probably have affected in the nutritional status, and consequently the body size, of the animals that were in juvenile phase during meagre periods.

Slaughter patterns of cattle

The overall slaughter pattern of cattle was analysed by recording frequencies of freshly fractured bone, cut marks and chop marks. 14 % of the cattle long bones were freshly broken, probably for marrow extraction. This figure may be too low, considering that a major portion of broken bone shaft fragments could not be identified to species. Especially metapodials, but also humeri, radii, femora and tibiae, were often broken for marrow extraction by craniocaudally splitting the whole bone, proximal or distal end, or both ends (Fig. 14). 37 % of the metapodial fragments were split this way.

A frontal bone carried chop marks around horncores, indicating removal of horns. Five mandibulae carried cut marks especially around the lateral side of the ascending ramus and below the articular process, and five mandibulae were broken or carried chop marks in the diastema. One hyoid bone carried cut marks, perhaps indicating the consumption of the tongue.

The vertebral column seems to have been split longitudinally, which can be seen in four vertebral fragment broken craniocaudally, and also divided to smaller units by mediolateral splitting, which was observed in eight vertebral fragments.

The scapula seems to have been removed by chopping near the glenoid, which was observed in 23 scapula fragments. This joint was also sometimes broken by chopping the proximal humerus, which can be seen in four humeral fragments. The elbow joint was also broken, indicated by 30 distal humeri, 22 proximal radii and 10 proximal ulnae with cut marks or pieces cut off.

Pelvis was often chopped to pieces, especially around the acetabulum. 14 pelvis fragments were chopped this way. Also 16 proximal femora were chopped from the proximal end. The knee joint was broken by chopping the distal femur (9 fragments with chop marks) or proximal tibia (3 fragments with chop marks).

Overall, the butchery marks indicate that cattle were probably butchered from the shoulder, elbow, hip, knee and ankle joints. The spine was craniocaudally split and cut into smaller sections, and horns or antlers were sometimes chopped off. Metapodials, and possibly other long bones, were also split for marrow extraction. The skin was removed, which left traces mainly on the mandible and ankles, because the skin is rather closely attached to bone in these areas (cf. Landon 1996:66-93).

Skeletal frequencies were analysed in order to discuss the meat trade in Tornio. Skeletal frequencies of cattle bones dating to different periods are presented as NISP, MNE and MAU (App. 10). MNE and MAU are presented because of the problems associated with NISP-based skeletal frequencies. Especially the fragmentation of the skull and greater number of bones in the skull and limb extremities brings about the risk of overrepresentation of these elements in NISP-based counts (e.g. Lyman 1994: 223-293; Stiner 2002). In general, the MNE and MAU figures in each period indicate that crania, upper limbs and extremities are more common than vertebrae and ribs. The lower frequencies of vertebrae and ribs are due to their poor identifiability from fragmentary material. In general, upper limb elements such as the scapula, humerus, radius and femur are the most common, but there is some variation that does not seem to be associated with any patterns of human behaviour. For instance, the frequencies of metatarsal and metacarpal bones, which are used often as raw material in crafts (e.g. Armitage 1982), were not bigger than those of other elements belonging to the limb extremities, such as tarsal and carpal bones. And, there was no conclusive evidence on development of meat cut trade in Tornio; the frequencies of low-utility body parts such as the cranium and the phalanges, were only a little lower than those of high-utility body parts.

In addition to human action, also differential preservation of skeletal parts often affects the skeletal frequencies of archaeological animal bone finds (Lyman 1994:235-258). Structural density of bone is one of the factors that affect bone survival and it is highly variable within the different elements of the vertebrate skeleton (Lyman 1994:235-258; Stiner 2002). Thus, the MNE figures of the skeletal elements were examined more closely by plotting the structural density of each bone element against its representation in the assemblage from Tornio (Fig. 15). I used the bone mineral densities of a wild bovid, the wildebeest (Connachaetes taurinus) bones measured by a CT scan by Lam et al. (1999), as a proxy for cattle skeleton (Table 5). There are, of course, differences in bone mineral density between species according to, for instance, the locomotor mode (Lyman 1994:239). However, Lam et al. (1999) have proven that intra-skeletal variation in bone mineral density follows a similar pattern within the artiodactyls. Moreover, the skeletal anatomy and locomotor modes of within the bovids are quite similar and the wildebeest has a roughly similar weight (ca. 140-260 kg) with early modern cattle in Tornio (estimated as ca. 160 kg).

The scatterplots of bone structural density and MNE's on cattle in different periods are presented in Figure 15. It seems that in general, the skeletal elements with higher mineral density are also those that are found in bigger frequencies from the archaeological cattle bone assemblage from Tornio. Correlations are statistically significant (Fig. 15), except for the period 1721-1800, which probably has to do with the small number of bones dating to that period. Or, at least, the lack of correlation is hardly due to uneven representation of slaughter and kitchen waste, as, for instance, the MNE of cranium is relatively high and the MNE of the pelvis relatively low in 1721-1800. Thus, the variation in cattle skeletal frequencies seems to be in large part caused by density-dependent survival of skeletal elements.

Overall, the skeletal frequencies of cattle indicate towards slaughtering of cattle in the town and there is no conclusive evidence of specialised meat cut trade. This notion is supported by the scarce historical data. In small-scale urban systems in northern Finland and Sweden, it was common that people bought animals alive from a butcher and these were then slaughtered at home, either by the butcher or the purchasers (e.g. Virkkunen 1953: 158; Vretemark 2001: 45-46). Hence, the skeletal frequencies of this kind of urban sites tend to contain equal proportions of edible parts and butchery waste (Vretemark 1997: 65; Landon 1996). In Tornio, there is no historical record on a similar system, and the first reference to a butcher is from the late eighteenth century (Mäntylä 1971: 602).

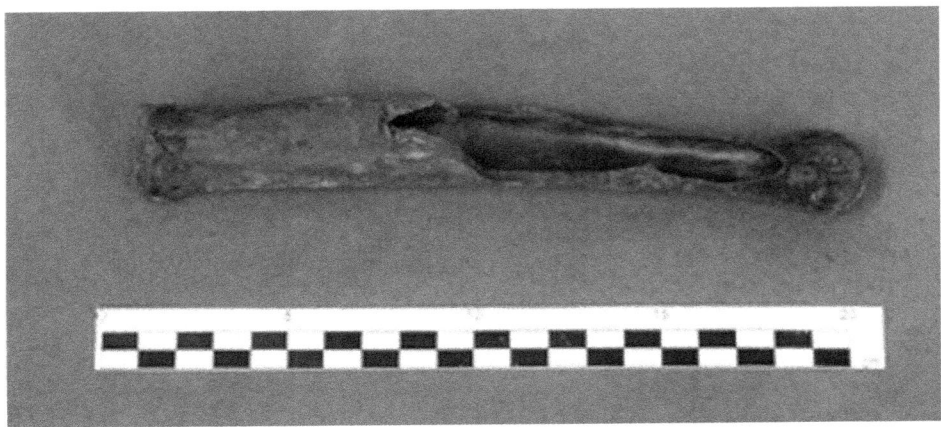

Figure 14. A cattle metatarsal with craniocaudally split distal end.

Table 5. Bone mineral densities of wildebeest skeletal elements according to Lam et al. (1999) and minimum numbers of elements (MNE) of cattle in different periods.

	wildebeest bone mineral density (Lam et al. 1999)	MNE 1621-1660	MNE 1650-1728	MNE 1721-1800
mandible (DN4)	0.62	16	4	8
atlas (AT1)	0.55	9	5	3
thoracic vertebrae (TH1)	0.38	2	1	1
scapula (SP1)	0.68	37	9	5
prox humerus (HU1)	0.32	14	5	3
dist humerus (HU5)	0.51	25	9	2
prox radius (RA1)	0.51	31	8	6
dist radius (RA5)	0.47	16	6	4
ulna (UL2)	0.73	31	8	6
prox metacarpal (MC1)	0.58	16	7	1
proximal phalanx (P11)	0.47	8	4	3
medial phalanx (P21)	0.47	8	4	2
distal phalanx (P31)	0.42	5	2	2
acetabulum (AC1)	0.64	26	7	2
prox femur (FE1)	0.41	24	4	6
dist femur (FE6)	0.38	18	3	
prox tibia (T11)	0.42	4		
dist tibia (T15)	0.48	20	8	2
talus (AS2)	0.77	26	9	3
prox metatarsal (MR1)	0.63	22	12	1

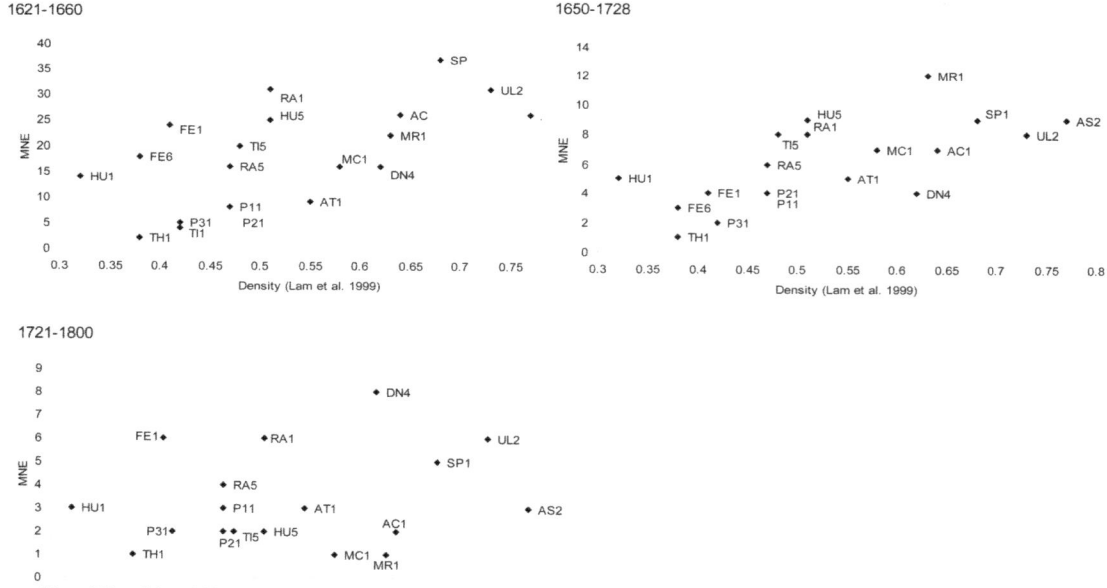

Figure 15. Scatterplots of bone mineral densities of wildebeest skeletal elements according to Lam et al. (1999) and minimum numbers of elements (MNE) of cattle in different periods. Correlations: 1621-1660 r = 0.619 p = 0.002, 1650-1728 r = 0.666 p = 0.001 1721-1800 r = 0.223 p = 0.186.

Sheep or goat

Age profile of sheep or goat

Sheep (*Ovis aries*) or goat (*Capra hircus*) age profiles were studied using dental eruption and wear and epiphyseal fusion. Majority of the anatomical elements are from animals older than three years according to epiphyseal fusion (Fig. 16). Also the combined data of tooth eruption and wear (Fig. 17, App. 11) indicates that mainly mature individuals were killed. A large part of the population had a M.W.S. over 36 points, and one individual had a M.W.S. of 45-49 points, which in Vretemark's (1997:88) material indicates to the age of over 6 years. The differences in the killoff pattern of sheep were compared using epiphyseal fusion data (Fig. 16); there are no significant differences between the periods in regard to the slaughtering age of sheep. The notch in the eighteenth century killoff pattern is most likely caused by the small sample size, only six bone fragments.

The age profiles indicate a dominance of wool production over meat production. It is also possible that sheep or goat were milked, although this was not commonly practiced in Finland (Wilmi 2003:178). Furthermore, the even sex ratio (see below) infers to dominance of wool production over milking. In late eighteenth century and early nineteenth century Oulu, younger sheep were slaughtered, probably for meat (Puputti 2007), but nothing similar can be seen in the animal bone material from Tornio.

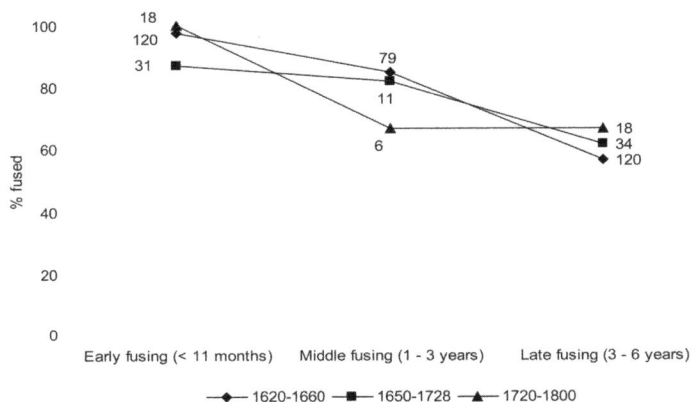

Figure 16. Proportions of fused epiphyses in each age group, divided by period. The early fusing elements are proximal scapula, acetabulum, proximal radius, distal humerus and proximal 1st and 2nd phalanges. The middle fusing elements are distal metapodials and distal tibia. Late fusing elements are femur, proximal calcaneum, proximal humerus, distal radius, proximal tibia, proximal ulna and vertebrae. Total numbers of anatomical elements are also presented.

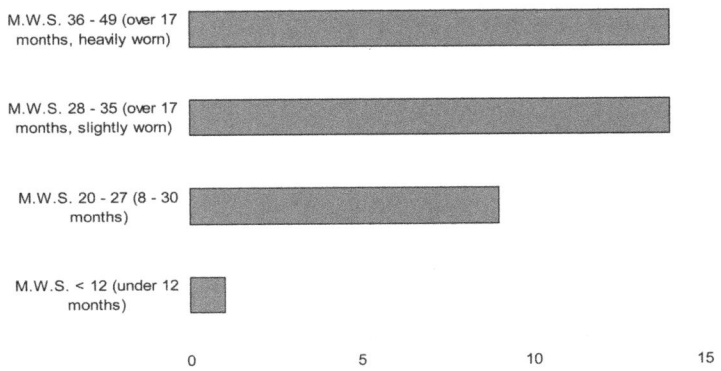

Figure 17. Combined data on tooth eruption and wear of ovicaprids.

Pelvic morphology and metacarpal measurements were used in sex assessment of sheep or goats. The criteria used for sex identification in pelvic fragments were from Boessneck (1969). In the bone material from Tornio, there were roughly equal numbers of male and female sheep or goat pelvises. The medial thickness of acetabulum was not examined in the case of sheep as it has proven unreliable in sex assessment of this species (Tourunen 2008:100). A total of 29 male and 27 female pelvises of sheep or goat were found, and these are presented in detail in Appendix 12.

In addition to pelvic morphology, also metacarpal measurements were examined. All the metacarpals included in metric study were identified as sheep, and thus, different dimensions of sheep and goat metacarpals do not affect the sex assessment. According to Davis's (2000) study on modern sheep, the most useful metacarpal measurements in sex assessment seem to be metacarpal slenderness plotted against greatest length, and metacarpal distal breadth plotted against shaft width.

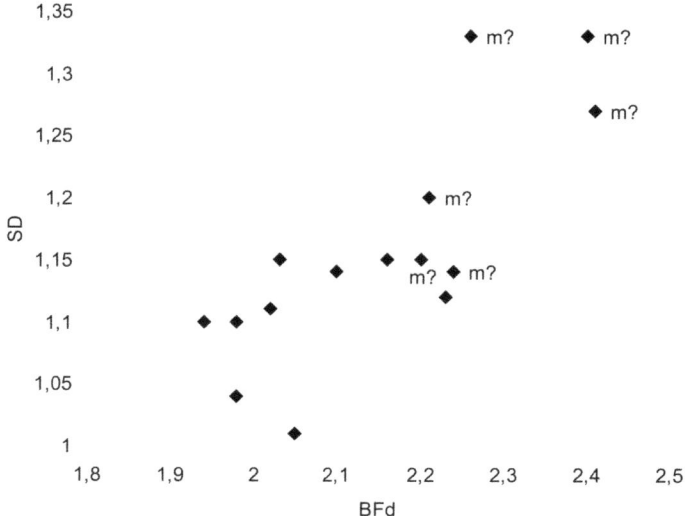

Figure 18. The scatterplot of the physiological distal breadth (measurement definition in Davis 1996) and shaft width (mm) of sheep metacarpals. m? = possible males.

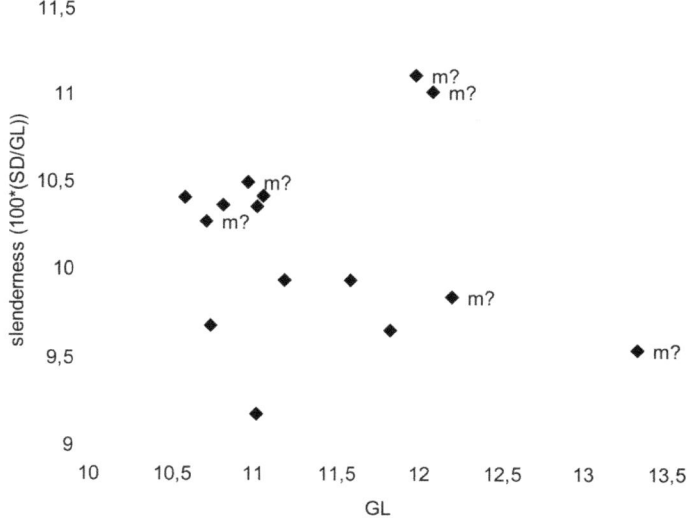

Figure 19. The scatterplot of the greatest lenght (mm) and slenderness index of sheep metacarpals. m? = possible males.

According to Davis (2000), metacarpal measurements are inferior to the plots of greatest lengths and shaft widths of tibia and humerus in sex assessment of sheep. However, as complete long bones except for metapodials were extremely rare in the assemblage, these plots could not be used in sex assessment of sheep. Moreover, the identification of sheep and goat was impossible of the majority of the bone fragments, while the metapodials were identifiable. Thus, the possible presence of two species does not confuse the sex assessment from metapodials as it might when using other long bones.

The scatterplot of metacarpal distal breadth against shaft width (Fig. 18) shows two relatively clear clusters of slenderer and stockier metacarpals, the latter of which may be male and are marked with 'm' in the figure. The metacarpal greatest length plotted against slenderness index (calculated as 100*(SD/GL) as defined by Davis 2000) (Fig. 19) does not show as clear clustering, but the majority of the individuals assessed as males based on Figure 18 seem to be rather stocky and relatively short. One individual has a particularly long and quite slender metacarpal. Although it is impossible to make definite conclusions based on only one specimen, the possibility of this individual being a castrated male cannot be excluded. Male sheep are sometimes castrated in order to improve the quality of the fleece and meat and reduce aggression (Davis 2000).

Size of sheep or goats

The data on postcranial skeletal measurements of sheep or goat bones are presented in Appendix 8. Again, the log ratio method was employed. The log ratios, divided by period, are presented in Appendix 9 and Figure 20. No significant differences in the size of sheep or goat were observable in the material.

The withers heights of sheep were estimated based on metacarpal greatest length (Table 6). I used Teichert's coefficient (von den Driesch & Boessneck 1974) in order to be able to compare the results with withers height estimations of early modern and medieval sheep in Sweden (Sten 1994). The resulting mean withers height, 55.7 cm, is similar to estimated mean withers heights of sheep in early modern Sweden, ca. 57 cm (Sten 1994:46).

The body masses of sheep were estimated with the regression equations for bovids by Scott (1990). As with cattle, distal humerus, distal tibia and proximal radius were chosen for body mass predictions because of their relative abundance in the bone assemblage from Tornio, as well as their low rates of estimation error. The results are presented in Table 7. The resulting body mass estimations for early modern sheep are very high, ca. 65 kg, and contradict the relatively small estimated withers heights. For instance, modern-day male Finnsheep weigh ca. 87 kg and females 60 kg, with withers height of ca. 72 cm for males and 65 cm for females (www.tiho-hannover.de). It is unclear why Scott's (1990) equations would severely overestimate the body masses of early modern sheep, as her data included also small species of bovids (Scott 1990). Again, the suitability of these equations to domestic sheep may be questioned.

Body mass estimations of archaeological sheep and historical data of live weight of sheep are rare, but medieval sheep in York seem to have weighed ca. 37 kg with withers height of ca. 59 cm (O'Connor 2003:179). With the use of O'Connor's (2003:179) regression equation, based on a sample of modern sheep radii, the mean body mass of early modern sheep in Tornio is ca. 32 kg. This estimation is more in accordance with the withers height estimations of sheep from Finland and Sweden during the early modern period. Hence, sheep in early modern Tornio seem to have been rather small, with withers height corresponding to sheep in early modern and medieval Sweden (Sten 1994). The body mass estimations, however, produced somewhat controversial results and no reliable conclusions can be made.

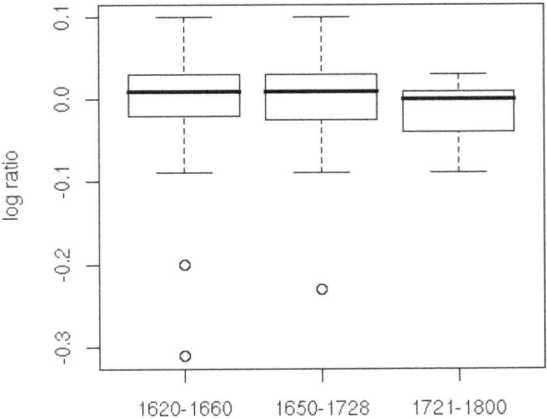

Figure 20. Comparison of log ratios of sheep or goat skeltal measuremens between the periods 1620-1660 (N= 253 ± 0.04), 1650-1728 (N= 51 ± 0.05) and 1721-1800 (N= 18 ± 0.03). Two-way ANOVA results are F = 0.9 p = .401.

Table 6. Withers height estimations of sheep based on metacarpal greatest lenght, estimated with the coefficient of Teichert (von den Driesch & Boessneck 1974).

Coefficient	N	Mean (cm)	Stand.dev.	Range (cm)
4.89 (Teichert)	16	55.7	3.6	51.7 – 65.1

Table 7. Body mass estimations of sheep or goat based on skeletal measurements, according to regression equations of Scott (1990) and coefficient of O'Connor (2003).

Measurement	N	Mean (kg)	Stand.dev.	Range (kg)
hum Bd	36	67.3	4	60.5 – 78.0
hum BT	33	61.4	3.8	51.8 – 70.4
tib Bd	56	64.2	3.2	57.2 – 73.2
rad Bp	34	65.7	3.9	56.5 – 72.8
all body mass estimations	159	64.6	4.2	51.8 – 78.0
rad Bd (O'Connor 2003)	19	32.3	2.4	28.1 – 38.4

Butchery patterns of sheep or goat

Butchery marks and evidence of marrow extraction were considerably rarer in ovicaprid than in cattle skeletons, which may be due to practical matters in handling the carcasses of smaller species, or, cultural practices. Overall, however, the butchery marks on the ovicaprid bones indicate to similar way of disarticulating the carcass as with cattle.

One back of the skull of a goat had the horns chopped off, and three fragments of axis vertebrae were chopped, which may have to do with the removal of the head. One cervical vertebra was mediolaterally split and one lumbar vertebra carried a cut mark on the vertebral body. One proximal scapula was chopped off, eight humeri had the distal epiphyses, or parts of distal epiphyses chopped off, five proximal ulnae were chopped off and one proximal radius attached to an ulna was chopped. One radius and one humerus had their diaphyses broken. Five pelvis fragments were broken around the acetabulum, and they also carried cutmarks around the acetabulum. Two caput femoris were chopped off, and one distal femur carried a chop mark on the medial side. Two proximal tibiae carried chop marks and one distal tibia carried a cut mark on the cranial side. Five shafts of tibiae were broken. Four metapodials were broken and one calcaneus carried a cut mark on the lateral side.

NISP, MNE and MAU figures of sheep or goat bones in different periods are presented in Appendix 13. All parts of the skeleton, i.e. cranium, axial skeleton, upper limbs and extremities are present in the assemblage in each period, although the small sample sizes in late seventeenth century and eighteenth century make comparison difficult. However, the upper limb bones as somewhat better represented than other body parts. The low representation of axial skeleton is probably a product of the poor identifiability of fragmentary vertebrae and ribs. The somewhat poorer representation of crania and limb extremities may be related to some degree to meat trade. This question was examined in more detail by comparing the bone mineral densities of sheep (Lyman 1994:240-248) and the MNE's of sheep bone finds from early seventeenth century Tornio (Table 8, Fig. 21). Only the finds from the early seventeenth century were plotted, because the sample sizes dating to later periods were very small. The correlation between bone densities and frequencies was not statistically significant ($r = 0.22$ p 0 0.19). In Figure 16, proximal metapodials, talus and mandible, all low-utility body parts, seem to have considerably "too low" representations in regard to their mineral densities. This further confirms the idea that low-utility skeletal elements of sheep are less represented in the early seventeenth century assemblage from Tornio than higher-utility elements. This may be an indication of sheep meat trade in the town, although it has to be kept in mind that low-utility parts were also present in the assemblage and thus at least some sheep or goats were slaughtered in the town. In late eighteenth century and early nineteenth century Oulu, the meaty body parts dominate the ovicaprid bone assemblages more clearly than in Tornio, indicating the development of meat trade (Hänninen 2008).

Table 8. Bone mineral densities of sheep skeletal elements according to Lyman (1994:240-248) and minimum numbers of elements (MNE) of sheep or goat in different periods.

	sheep bone mineral density (Lyman 1994:240-247)	MNE 1621-1660	MNE 1650-1728	MNE 1721-1800
mandible (DN4)	0.57	13	3	3
atlas (AT1)	0.13		1	1
thoracic vertebrae (TH1)	0.24	2	1	1
scapula (SP1)	0.36	15	3	2
prox humerus (HU1)	0.24	8		
dist humerus (HU5)	0.39	31	3	2
prox radius (RA1)	0.42	22	8	2
dist radius (RA5)	0.43	11	3	1
ulna (UL2)	0.45	21	2	3
prox metacarpal (MC1)	0.56	19	3	
proximal phalanx (P11)	0.36	6	1	1
medial phalanx (P21)	0.28	3	2	
distal phalanx (P31)	0.25	2	2	
acetabulum (AC1)	0.27	20	4	1
prox femur (FE1)	0.41	24	4	2
dist femur (FE6)	0.28	10	3	1
prox tibia (T11)	0.3	17	2	1
dist tibia (T15)	0.5	38	6	2
talus (AS2)	0.59	16	5	
prox metatarsal (MR1)	0.55	16	4	

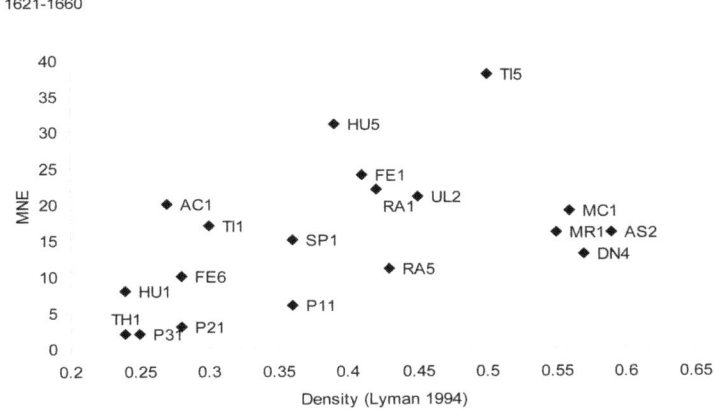

Figure 21. Scatterplot of bone mineral densities of sheep skeletal elements according to Lyman (1994:240-248) and minimum numbers of elements (MNE) of sheep or goat.

Pig

Age profile of pigs

Pig (*Sus scrofa domesticus*) bones were aged using the same methods as above, i.e. epiphyseal fusion and dental information (App. 14, Figs. 22 and 23). Pig bones were less frequent finds than bones of cattle and sheep, and thus, the comparison between periods is based on very small sample sizes, probably causing the fluctuations. Overall, it seems that pigs were slaughtered for meat at an early stage after reaching their adult body size. This seems to have been between the age of one and two years, based on ageing data extracted from modern pig populations. In early modern pigs, the dental eruption dates and epiphyseal fusion dates may have been somewhat later, indicating slightly older age for slaughtering. Slaughtering pigs at an early age is common in archaeological animal bone assemblages in Finland and Sweden, as these animals were kept mainly for meat production (Vretemark 1997; Tourunen 2008).

Sexing pig bones

Canine morphology (Mayer & Brisbin 1988) was used in sexing pig mandibulae (App. 15). A total of nine mandibulae were assessed as belonging to males and seven to females. Although the sample size is rather small, it seems that both sexes were represented fairly equally.

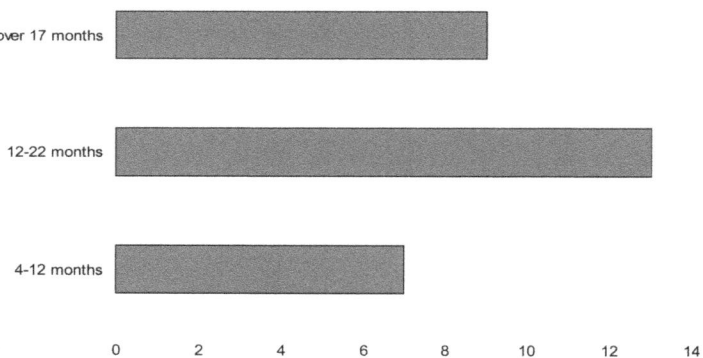

Figure 22. Combined data on tooth eruption and wear of pigs.

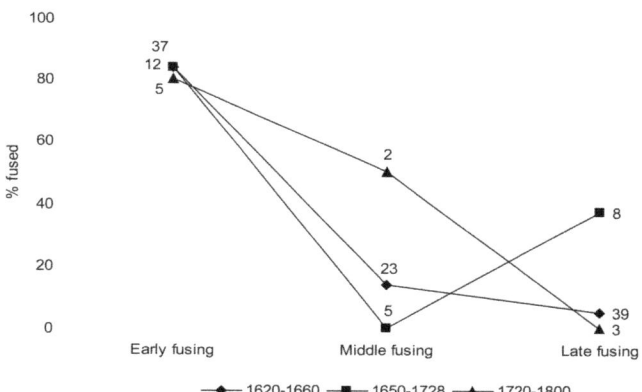

Figure 23. Proportions of fused epiphyses in each age group, divided by period. The early fusing elements are proximal scapula, acetabulum, proximal radius, distal humerus and proximal 1st and 2nd phalanges. The middle fusing elements are distal metapodials and distal tibia. Late fusing elements are femur, proximal calcaneum, proximal humerus, distal radius, proximal tibia, proximal ulna and vertebrae. Total numbers of anatomical elements are also presented.

The size of pigs

Pig bone sample from early modern Tornio is somewhat smaller than the samples of cattle and sheep or goat bones, which complicates the estimation of the size of pigs. Withers heights could not be estimated because there were no complete long bones in the assemblage. Body mass was estimated from distal humerus and distal tibia with Scott's (1990) equations for suids (Table 9). Mean of all pig body mass estimations is 78 kg, but the range and variation in weight predictions are considerable. The maximum distal breadth of humerus gives the largest estimations, up to 95 kg, whereas the breadth of the trochlea gives the smallest body weights, from 62 kg to 74 kg. Due to the lack of comparative data, it is difficult to assess the reliability of the body mass predictions for pigs in early modern Tornio, but they seem to have been considerably smaller than modern landraces: e.g., the males and females of the Finnish landrace weigh 250 and 200 kg, respectively (www.tiho-hannover.de). Pig bone finds from different periods were compared with log ratio-method (App. 9), but there were no statistically significant differences. The small sample sizes, merely 3 and 2 bone fragments from the late seventeenth century and eighteenth century, respectively, did not allow any significant conclusions (two-way ANOVA results were $F = 2.2$ $p = .130$).

Body part presentation of pigs

Butchery marks and evidence of marrow extraction were rather scarce in pig skeletons. Definite conclusions about the partitioning of pig carcasses cannot be made based on such scanty evidence, but it seems that also some pig bones were broken for marrow extraction.

Two cranial fragments, a frontal bone and the lateral side of a caudal mandible, carried cut marks perhaps indicating the removal of skin and/or jowl meat (cf. Landon 1996:70). Three distal humeri were carried chop marks and one humeral diaphysis was broken. One ulna carried a cut mark and the proximal ends of three ulnae were chopped off. One distal femur was chopped off and one pubic bone carried cut marks. In addition, one metapodial was broken.

NISP, MNE and MAU figures skeletal elements of pig dating to different periods are presented in Appendix 16. According to the both NISP and MNE figures, crania are the most abundant skeletal elements in the assemblage, while all skeletal elements are represented. Again, the lower MNE figure of elements belonging to the axial skeleton is probably caused by the poor identifiability of fragmented ribs and vertebrae to species. The fact that both low and hig utility skeletal elements are present indicates that also pigs were slaughtered and consumed in the town, while the comparison between periods is difficult because of the extremely small sample sizes in the late seventeenth century and eighteenth century.

Table 9. Body mass estimations of pigs based on skeletal measurements, according to regression equations of Scott (1990).

Measurement	N	Mean (kg)	Stand.dev.	Range (kg)
hum BT	15	69.0	2.9	61.5 – 73.5
hum Bd	13	89.5	3.8	81.2 – 94.6
tib Bd	4	76.0	5.7	69.1 – 83.0
all body mass estimations	32	78.2	10.4	61.5 – 94.6

Reindeer

Age profile of reindeer

The age of the reindeer (*Rangifer tarandus*) in early modern Tornio was analysed based on epiphysial fusion (Hufthammer 1995). Reindeer are included here in domestic animals, although it is acknowledged that both subspecies, the semi-domesticated reindeer (*R.t.tarandus*) and wild forest reindeer (*R.t.fennicus*), were probably present in Tornio (Puputti & Niskanen 2009). The bones are analysed as an entity, as the numbers of reliable subspecies identifications were rather small. Moreover, only fused bones were possible to identify to subspecies and thus comparing epiphysial fusion between the subspecies was not possible.

Majority of the reindeer have reached the age of over three years (Fig. 24). Thus, it is probable that mainly old draft reindeer were slaughtered for meat, and/or that calves of wild forest reindeer were not generally hunted. This kind of age distribution is typical also to reindeer bone assemblages in marketplaces in Lapland (Lahti 2006:291). Also in Oulu, where majority of the reindeer bones qualified for subspecies assessment were wild forest reindeer, old individuals dominate the assemblage (Puputti 2007). None of the unfused early-fusing elements in the materials from Tornio and Oulu were tuber scapulae which fuse at the age of less than six months (Hufthammer 1995; Puputti 2007), and thus it seems that the meat of semi-domesticated reindeer and wild forest reindeer calves was not consumed. In regard to wild forest reindeer, adults animals may have been preferred over calves, or, these animals were generally not hunted during the first six months after the calving season, i.e., in from late April to early December (Nieminen 1994:74).

The size of reindeer

Wild forest reindeer have slightly larger average postcranial skeletal dimensions (Puputti & Niskanen 2008), and longer limbs as an adaptation to the thick snow cover in forests during the winter (Nieminen & Helle 1980), which complicates the estimation of the withers height of archaeological reindeer. The body masses of archaeological reindeer from early modern Tornio were estimated with a regression equation created for semi-domesticated and wild forest reindeer (Puputti & Niskanen 2008). The resulting mean weight of early modern reindeer, 64-165 kg, corresponds well with the weights of modern reindeer females and males (Nieminen & Petersson 1990). Also the original skeletal dimensions of the archaeological reindeer bones correspond well with modern reindeer. The size of reindeer in different periods was compared with the log ratio method (App. 9). The standard is extracted from a sample of modern reindeer from the University of Oulu Zoological Museum (see Puputti & Niskanen 2008 for details), and there were no significant differences between periods. But, again, the sample sizes from the late seventeenth century and eighteenth century are very small, only three and six individuals in the late seventeenth century and eighteenth century, respectively. The two-way ANOVA results were $F = 1.5$ $p = .246$.

Body part presentation of reindeer

In all, the butchery marks on reindeer skeletons imply to similar butchery patterns with cattle. One fragment of chopped reindeer antler was found, perhaps evidence of using antler as raw material for carving objects or cooking glue. A short-lived glue factory that used reindeer antler as raw material was established nearby Tornio in the late eighteenth century (Mäntylä 1971:380-381), and this kind

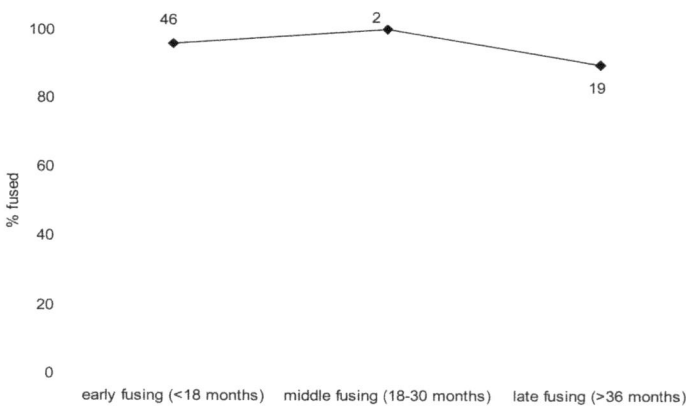

Figure 24. Proportions fused epiphyses in each age group, divided by period. The early fusing elements are proximal scapula, acetabulum, proximal radius, distal humerus and proximal 1st and 2nd phalanges. The middle fusing elements are distal metapodials and distal tibia. Late fusing elements are femur, proximal calcaneum, proximal humerus, distal radius, proximal tibia, proximal ulna and vertebrae. Total numbers of anatomical elements are also presented.

Table 10. Bone mineral densities of skeletal elements according to Lam et al. (1999) and minimum numbers of elements (MNE) of reindeer in the early seventeenth century.

	reindeer bone mineral density (Lam et al. 1999)	MNE 1621-1660
mandible (DN4)	0.67	2
scapula (SP1)	0.66	1
prox humerus (HU1)	0.26	1
dist humerus (HU5)	0.48	13
prox radius (RA1)	0.53	1
ulna (UL2)	0.68	1
proximal phalanx (P11)	0.48	1
medial phalanx (P21)	0.49	1
distal phalanx (P31)	0.48	1
acetabulum (AC1)	0.64	2
prox femur (FE1)	0.39	2
dist femur (FE6)	0.32	2
dist tibia (TI5)	0.39	1
prox metatarsal (MR1)	0.58	3

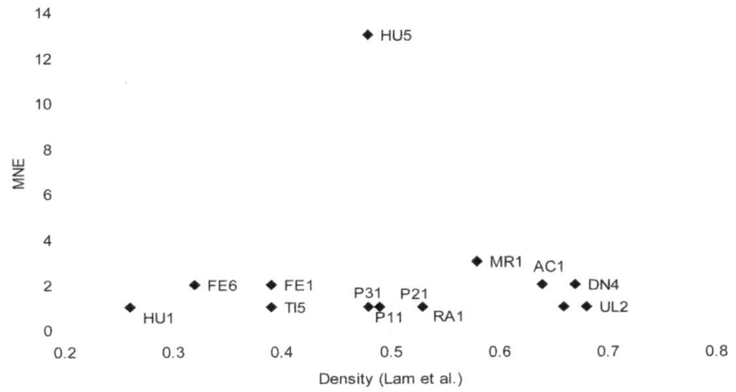

Figure 25. Scatterplot of bone mineral densities according to Lam et al. (1999) and minimum numbers of elements (MNE) of reindeer.

of activities may have been practices also previously. Two proximal scapulae were chopped from the medial side and three distal humeri were chopped. In addition, eight humeral shafts and one radius were broken. One fragment of pelvis had chop marks around the acetabulum, and one distal humerus had chop marks in the intercondylar fossa. One proximal femur had cut marks in the femoral head, which perhaps indicates a more delicate way of disarticulating the hip joint than the crude chopping of several cattle pelvises and proximal femora indicate. Additionally, four femoral shafts and one metatarsal were broken. The metatarsal also carried cut marks on the distal end, which may be related to the removal of the skin.

NISP, MNE and MAU figures of reindeer in different periods are presented in Appendix 17. According to the figures, elements belonging to upper limbs, especially the humerus, are the most abundant skeletal elements in the assemblage, while all skeletal elements apart from the axial skeleton are represented. Quality meat parts are slightly more common than low-quality parts. Especially the high representation of the humerus in the early seventeenth century merits attention. Although this may have simply to do with small sample size and poor identifiability of small cranial fragments, it is also possible that some reindeer meat cuts were imported to Tornio. Density-dependent survival of reindeer bones was examined by plotting the MNE's of the bone finds and the bone mineral densities of reindeer (Rangifer tarandus) (Lam et al. 1999) (Fig. 25, Table 10). This figure clearly shows that the frequency of distal humerus is far "too high" in regard to its mineral density, which further indicates towards selective import of reindeer skeletal parts. According to Lahti (2006:290), the possibility of exporting quality meat cuts, especially those containing the humerus, from the market places in Lapland, where the Sámi traded with merchants from

Tornio, cannot be excluded. Again, the absence of axial skeletal elements may be caused by the poor identifiability of fragmented ribs and vertebrae to species. The presence of all body parts, including the cranium, however, indicates that some reindeer were also slaughtered and consumed in the town.

Horse

The horse (*Equus caballus*) bone finds and their provenance are discussed in more detail in chapter 4. All the horse bone finds were fused, which indicates that horses reached an old age (over 15 months based on metapodials, over nine months based on phalanges and over three years based on the calcaneum) (Silver 1969; Barone 1999). According to the historical record, horses were quite common in early modern Tornio. Horses were used in fieldwork in northern Finland (Virrankoski 1973:197), but it seems that in the Tornio River Valley, horses were mainly used in transportation (Teerijoki 1993:91).

The withers height of horse was estimated according to the coefficients provided by von den Driesch & Boessneck (1974), based on metapodials. The horse withers height estimations from a metacarpal (greatest length 208 mm) and a metatarsal (greatest length 242 mm), 128 cm and 133 cm, respectively, support the notion that the Finnish horses of the period were very small. Historical military records and museum collections of collars reveal that the estimated average withers height of Finnish horses was less than 120 cm in the seventeenth century and about 128 cm in the eighteenth century (Vuorela 1975:196), whereas horses used by the military as cavalry mounts were considerably taller. The minimum accepted height was 139 cm during the last decades of Swedish rule (Screen 2007). However, the body proportions of a given horse breed or type can considerably affect the relationship between metapodial length and withers height, and conversion coefficients based on different breeds can therefore produce widely differing withers height estimations (Niskanen 2008:66-67).

Domestic hen

12 fragments of domestic hen (*Gallus domesticus*) bones were found among the animal bone assemblage. The estimation of numbers of hen in early modern Tornio is difficult as they were not taxed (Virrankoski 1973:242). According to the probate inventories, however, it seems that hen became more common during the eighteenth century while they are rarely mentioned in earlier probate inventories (OMA BIa:1–9.).

5.3. Hunting wild species

Seals

All but two excavation areas (Areas 3 and 4) in Tornio produced small quantities of seal bones. Two bone fragments were identified as grey seal (*Halichoerus grypus*), and seven bone fragments as ringed seal (*Phoca hispida*). The percentages of seal bones in the assemblages varied from zero to 3.4 % in different excavation areas. The small but constant representation of seal bones implies that sealing seems to have been a regular but not primary activity.

Furthermore, according to tax records, the role of sealing has been of lesser importance in the northern Ostrobothnian parishes than in the southern and middle Ostrobothnia (Luukko 1954:438-439; Kvist 1990), the emphasis being on livestock husbandry, salmon fishing and inland hunting (Kvist 1990; Ylimaunu, J. 1996). Seal meat, skins and blubber were fairly important export products in Tornio in the early modern period (Mäntylä 1971: 50; Virrankoski 1973: 277). The historical records on sealing during the seventeenth and eighteenth centuries, however, are not as abundant as during the medieval period, as blubber and seal catch of long-distance trips were not taxed after the beginning of the seventeenth century (Ylimaunu, J. 1996: 181-182). Blubber was the most valuable product obtained from seals, but skins and meat were also sought after (Ylimaunu, J. 2000: 332-341).

Age profile based on epiphysial fusion (fusion dates from Storå 2001) indicates that the emphasis has probably been on adult or juvenile seal hunting as no bones belonging to yearlings were identified. The age profiles are related to the seasonality of sealing. Generally speaking, seals were hunted mainly in the fall and during the winter/early spring since in the summer the valuable fat layer is at its thinnest. Fall was typically the season of net hunting, whereas during the winter and early spring, the seals were taken from their breeding grounds on the ice (Luukko 1954: 439-441; Kvist 1987: 48-49). The spring sealing trips generally produced quantities of valuable seal pup furs (Virrankoski 1973: 277; Ylimaunu, J. 2000: 102). Therefore, the age distribution of the urban archaeological bone finds from Tornio, showing no seal pup bones, may represent mainly the catch of the net hunting of the fall. According to the tax records, the typical sealing method in Tornio area in the 16th century was net hunting during the fall, whereas in the southern and middle Ostrobothnia, the sealing trips to the ice were more typical (Virrankoski 1973: 277; Ylimaunu, J. 1996: 196; Ylimaunu, J. 2000: 102).

All skeletal parts were identified from the urban archaeological bone assemblages, including cranium fragments and upper limb bones (App. 4; Puputti in press b). The presence of the axial skeleton and upper limb bones indicates that whole seals were brought to the towns and therefore at least some of the sealing probably occurred in the vicinity of the towns. The seal catch of the long-distance trips was usually processed in the kill site and only blubber and skins were brought home, as the seals caught from nearby could be transported home as complete carcasses (Talve 1996:174).

Arctic hare

Bones of arctic hares (*Lepus timidus*) were present in all excavation areas and they were also still hunted during the eighteenth century, when bones of other game animals become scarcer. All the body parts of arctic hares were represented in the assemblage and also one bone of juvenile hare was found. The estimation of hunting season based on this find, however, is difficult as hares can give birth from early spring to late autumn (Nummi 1997). Generally, however, fur-bearing species, such as squirrels, red foxes and arctic hares were hunted during the late autumn and winter because of the quality of the winter fur (Vuorela 1975). Arctic hare furs were not very valuable in fur trade (Luukko 1954:387), but nevertheless, they were exported to Stockholm (Mäntylä 1971:70-71). Given the low value of the hare fur, and the fact that whole hare carcasses were brought to the town in stead of merely furs, it is likely that hares were mostly sought after for their meat. There are variable attitudes toward hare meat in the Finnish folklore (Talve 1973:43), but it seems that hare meat has been consumed in northern Finland (Vahtola 1997:127).

Red squirrel

Merely four bone fragments of red squirrel (*Sciurus vulgaris*) were found. This may partly be due to the lack of sieving and subsequent loss of bones of small species. However, the bones of small anatids and hazel grouse, only a little larger than those of red squirrels, are present in larger quantities and thus, it is likely that squirrels were not taken in considerable numbers by the townspeople. Fur of red squirrel was valuable in comparison with that of arctic hare and they were taken in large numbers during the 16th century (Luukko 1954:384-387). Some red squirrel furs were exported from Tornio in the seventeenth century (Mäntylä 1971:70-71), but the quantities are not comparable with those from the late medieval period when fur trade was at its most active (Luukko 1954:384-386). There were not cut marks on squirrel bones, or other indications of squirrels being eaten, but squirrel meat is known to have been sometimes consumed in traditional Finnish culture (Talve 1973:43).

Red fox

Red fox (*Vulpes vulpes*) were not abundant in the bone material as only 22 bone fragments identified as red fox were found. Also one unidentified canid bone was found, and it may derive from red fox, as well as from domestic dog, wolf or arctic fox. Red fox fur was very valuable, although the price somewhat depended on the colour of the fur (Luukko 1954:392). Considering the general aversion of carnivore meat in traditional Finnish culture (Talve 1973) and the fact that no evidence of butchery were found on the fox bones, it seems likely that these animals were mostly sought after for their fur. In the early seventeenth century Tornio, fox furs are mentioned among the trade goods (Mäntylä 1971:70-71) and it is probable that in addition to hunting foxes themselves, the tradesmen bought quantities of fox furs from the farmers and the Sámi.

Brown bear

The bear (*Ursus arctos*) claws found among the faunal material constitute a somewhat anomalous find as they probably originate from a foundation deposit (chapter 4). Other parts of bear skeletons were not found, and the nails may have arrived to the town attached to a fur. It is possible, though, that bears were hunted by the townspeople. Bear furs were valuable (Luukko 1954:396) and from the mid seventeenth century on, bounties were also offered of bears that were considered dangerous to cattle (Klemettinen 2002:145). Bears may have been disarticulated and skinned on the kill site, or, their remains may have been handled outside town area. There are accounts from eastern Finland on ritual treatment of bear bones: the skulls were hung on a pine tree and the postcranial bones were buried nearby (Klemettinen 2002:137-139). Ritual bear burials were also practiced by the Sami in the period from the seventeenth century to the nineteenth century; these have been found in present-day northern Sweden and Norway (Zachrisson & Iregren 1974).

Cervids

Wild forest reindeer bones are discussed together with semi-domesticated reindeer bones in the previous chapter. In addition to wild forest reindeer bones, one fused proximal humerus of elk (*Alces alces*) was found from Tornio. Large cervids may have been hunted more intensively than it appears based on the animal bone assemblage, as at least the initial butchery of these heavy animals must have occurred near the kill site and perhaps only meat cuts have been brought to the town. In rural northern Finland, wild forest reindeer and elk were hunted during the 16th and seventeenth centuries (Luukko 1954:396; Virrankoski 1973:272). Large cervids were generally hunted during the autumn, when the fat content of these animals is at its peak, or they were chased in the late winter on the crusted snow that hinders the movement of the animal but enables the skiing hunter to move rapidly (Vuorela 1975:57). As discussed above (Chapter 5.2.4.), mainly mature reindeer were taken which may imply to hunting during the wintertime.

Gallinaceous birds

Bones of wild gallinaceous birds were very common in the archaeological animal bone material, capercaillie (*Tetrao urogallus*) bones being the most abundant. Also black grouse (*Tetrao tetrix*), willow grouse (*Lagopus lagopus*) and hazel grouse (*Bonasa bonasia*) were taken. Willow grouse and rock ptarmigan (*Lagopus mutus*) are rather difficult to identify to species and thus, some bone fragments were identified only as *Lagopus* sp. However, as the habitat of the rock ptarmigan consists of treeless tundra and mountain area (Lokki & Palmgren 1992), it is probable that the birds caught nearby Tornio were willow grouse. Gallinaceous bird hunting also continued during the eighteenth century when catching waterfowl was becoming rare. All the wild gallicaneous bird species encountered in the animal bone assemblage are non-migratory species which range in forested parts of Finland (Lokki & Palmgren 1992). These birds could be hunted throughout the year, although traditionally they were caught mainly during the late autumn and winter, or during the spring display (Vuorela 1975:63-73). Five bones of young capercaillies were found from the seventeenth century assemblage, which indicates to capercaillie catching during the summer months. In addition to meat, gallinaceous birds may have been caught for their feathers, which were exported from Tornio during the early modern period (Mäntylä 1971:70-71).

Waterfowl

Waterfowl were hunted mainly during the seventeenth century, whereas among the eighteenth century finds their bones are rare. Whooper swans (*Cygnus cygnus*), greylag geese (*Anser anser*), lesser white-fronted geese (*A. erythropus*), bean geese (*A. fabalis*), Anas species, goosanders (*Mergus merganser*), red-breasted merganser (*M. serrator*), greater scaups (*Aythya marila*), tufted ducks (*Aythya fuligula*) and long-tailed ducks (*Clangula hyemalis*) were hunted. Waterfowl, on the contrary to gallinaceous birds, cannot be hunted year round, as they migrate to the south for the winter months. Thus, catching waterfowl is restricted only to spring, summer and autumn. Whooper swans, greylag geese, bean geese, Anas species, mergansers, greater scaups, tufted ducks and long-tailed ducks are migratory species which nest in marine, freshwater and marsh environments in northern Finland, while lesser white-fronted goose nests in the northernmost Lapland and migrates via the coast of the Gulf of Bothnia (Lokki & Palmgren 1992). Thus, this species may have been caught only during its migration period, in May or in August and September (Lokki & Palmgren 1992).

6. Discussion

6.1. Diet differences, social status and identity

People often articulate their distinctiveness through food choices, and this is why food preferences are often strongly interwoven with identities (Counihan 1999:7; Scholliers 2001). Moreover, sharing and acquiring food and suitable foods for different social groups are often socially controlled (Counihan 1999:10; Apo 2001:170-185; Stark-Arola 2001). The study of archaeological food remains, such as animal bones, thus yields information on social dynamics, power relations and identity construction.

Early modern Tornio was inhabited by people of various origins, especially during the first decades after the town's foundation (Mäntylä 1971:33-36). Thus, the emerging urban identities were constructed in the interaction between people of various origins, and foodways may have played a role in manifesting these identities. In addition to geographic origin, also social and economic status may have been components in the construction of identity. According to the taxes paid by the merchants, wealth was rather unevenly distributed among the townsfolk, as the majority of the trade concentrated in the hands of few rich merchants (Mäntylä 1971:114). Apparently, wealth also ran in the family, as the persons paying the highest taxes seem to be the same merchants, or their sons or widows, year after year (Mäntylä 1971:123). Wealth also seems to correlate with the location of the house. Richest people preferred to live along Rantakatu Street, while Takakatu Street (present-day Keskikatu) and Third Street were populated by the less affluent. This pattern is documented at least in the late seventeenth century and continued in the eighteenth century (Mäntylä 1971:125-126, 244). Thus, the distribution of social groups and wealth in Tornio are known on a general level while there is a lack of knowledge on how social and economic differences were actually displayed in everyday life.

The comparison of animal bone assemblages from different excavation areas showed that there were indeed differences in the characteristics of the animal bone assemblages. Especially the share of wild animal bones in the food waste from a plot (Figs. 6 and 7) varied in the town. There are two probable explanations for the varying ratio of wild animal bones. First, hunting may have been associated with low status. In central Europe, hunting is strongly associated with upper classes (e.g. Morris 2005; Sykes 2005), whereas in northern Finland, its social setting was completely different. At least in the late eighteenth century Oulu, reliance on hunting probably correlates with a low status (Ducey 2009). Moreover, during the seventeenth century, forests were shared by villages, not divided between individual landowners and thus, in principle, everyone could hunt regardless of wealth (Virrankoski 1973). In Tornio, however, it is unclear whether such correlation can be found. Hunting was relatively common among all townspeople and clearly formed an essential part of the everyday livelihood of people in the seventeenth century, and thus it is possible that it was not considered a lower-status activity.

It is possible, though, that hunting was associated with the geographic origin of Tornio. The proportion of wild animal bones is generally merely one or two per cents in early modern Southern Finland and Central Sweden, and the bones mostly belong to a few species of small mammals and waterfowl (Vretemark 2003; Tourunen 2008). Even in Oulu and Pietarsaari, both located a few hundred kilometres south of Tornio on the coast of the Gulf of Bothnia, the proportion of wild animal bones are generally less than 10 %, and the majority of the wild animal bones belong to arctic hares and gallinaceous birds (Puputti 2005, 2007a, 2007b, 2008). Hence, a lower share of wild animal bones and low species diversity in a seventeenth century animal bone assemblage from a given plot in Tornio could, in principle, indicate a more southern origin of the people, while a higher proportion of wild animal bones could imply local origin.

There were also differences in the diversity of domestic species (App. 1-3, Fig. 6) among the excavation areas. Probate inventories show clear differences in animal property between the affluent and the poor, the latter having fewer animals than the former. For instance, Johan Eriksson Giers, a church official with an estimated fortune of 265:2 copper dalers, owned one cow, while Johan Sigfredsson, a member of the city administrative court with an estimated fortune of 46.149:13 copper dalers, had a twelve-year old horse, six cows, four heifers, 10 chicken and 12 reindeer (OMA BIa:1). These and other probate inventories indicate that, generally speaking, a diversity of domestic species seems to be connected to wealth. The same phenomenon is also evident in the probate inventories from the late eighteenth and the early nineteenth century Oulu (Ducey 2009). High diversities of domestic species were not, however, observed in the same excavation areas in Tornio throughout the studied period, nor were they consistent with the indications from the comparison of the ratios of wild animal bones (Fig. 6).

Thus, the interpretation of the observed differences in the framework of social identity manifestation was not without problems. The putative indications of identity on a given excavation area were regularly contradictory and pointed to different directions. Although there were clear differences between the plots, the differences were by no means clearly explicable by a single factor such as wealth or geographic origin, or combinations of such factors, nor did they remain constant in the same plot throughout the studied period.

There are several possible reasons for the incoherent results. The results are related to the complex relationship between social identity and material culture, and the fact that the ethnic, national or geographically distinct cultural groups present in the historical documents are not necessarily reflected in the archaeological material (cf. Jones 1999). It has been suggested that high status and wealth were not straightforwardly manifested in the display of expensive ceramics, foodstuffs or townhouse architecture in early modern Tornio (Nurmi 2004; Herva & Ylimaunu 2005; Puputti 2006; Ylimaunu 2007), and the applicability of the sosioeconomic status model in the small towns of Finland and Sweden has been criticised (Rosen 1999; Nurmi 2004; Herva & Ylimaunu 2005). This model, based on the idea of consumer choice, suggests that wealth and high social status are manifested in purchasing expensive commodities and luxury items which can then be detected in the archaeological material (e.g. Hållans & Andersson 1992). However, it may have been that manifesting status was not meaningful in the small towns of northern Finland and Sweden, and it was not necessary where everybody knew each other (Rosen 1999; Nurmi 2004). For instance, the availability of certain foods and materials, and the general characteristics of the market system may affect the food choices and archaeological animal bone assemblages far more than the need to express social distinctiveness through food preferences (Landon 1996:115).

Moreover, the focus of this study, i.e. the comparison of whole plots, is not necessarily sufficient to these kinds of questions. This is because early modern households consisted of individuals with several different social identities and statuses, such as servants, widows and children. The number of these people is difficult to evaluate based on historical records, which mostly concern the activities of wealthy adult males (Vainio-Korhonen 1999), but they have been present and have contributed to the animal bone assemblages found from the town. Social position, for instance wealth, marital status or occupation, may have strongly affected the food choices and options of a person (e.g. Counihan 1999:10; Apo 2001:170-185). Thus, the animal bone finds from a given excavation area do not represent a single social group, but a number of social groups with possibly differing, socially controlled diets.

6.2. Living with animals

The presence of animals has had a profound impact on people's experience of the world in the past, and the interaction with different animal species has been an important part of everyday life (Whittle 2003:78; Armstrong Oma 2007:73-75). In the remaining chapters, I will rely mainly on the ideas of relational worldview (e.g. Bird-David 1999) and Ingold's concept of dwelling (e.g. Ingold 2000; 2006) in my interpretation of human-animal relationships and the changes that took place in these relationships in early modern Tornio. I will argue that the human-animal interaction that can be traced through urban archaeological animal bone material can be used in re-interpreting early modern urban human-animal and human-environmental relationships that have so far been considered mainly in an economical frame of reference.

The idea of relational perspective has been used in anthropological theoretical discussion to reconsider a group of cultural phenomena that have been labelled 'animistic' in earlier anthropological literature (Viveiros de Castro 1998; Bird-David 1999). According to Bird-David (1999:67), animism has been defined as the attribution of life, divinity or soul to such phenomena where there is none according to the Western scientific interpretation. She suggests that instead of treating these kinds of ideas as misapprehensions of the true nature of things, we should consider them as an indication of a different perspective on the world (Bird-David 1999:68). According to Ingold (2006:9-11), the so-called animists or relationists do not differ so much from us in what they think of as being alive. Rather, the perception of what is life is different. To them life is not an attribute of certain things but a way of participating in the world and in the continuous recreation of the world (Ingold 2006:9-11). In this way, relational perspective means focusing primarily on the relatedness of things; shared relationships and how things react and relate to each other are important (Bird-David 1999:69-77).

This relational perspective has a profound effect on the outlook we have on the relationships between people and animals. Rather than seeing animals as mere objects, we can focus on the relatedness of humans and animals; how people see animals and act towards them, and how people think animals see people and react to them (Viveiros de Castro 1998:477-478; Ingold 2000:51). This does not mean that animals would necessarily be regarded as similar to human persons, but rather, for instance, as goose-persons (cf. Ingold 2000:50; but see Willerslev 2007:74-75). These goose-persons may have same qualities as human persons, for instance consciousness or intentionality, but they interact with people and other geese in their own way (Viveiros de Castro 1998:478; Ingold 2000:50-51). However, relational perspective does not necessarily apply to all animal species or all the individuals of the same species (Viveiros de Castro 1998:471).

According to Ingold (2000; 2006), relational worldview suggests that the relationships between animals and humans can be understood in terms of their involvement with the world and each other; this is what he calls 'dwelling' in the environment (e.g. Ingold 2000:43-52). The idea of dwelling is rooted in phenomenological philosophy, which stresses the importance of lived experience in the world (Thomas 1998:19-20). According to phenomenological philosophers, people gain information on the world and give meaning to things through their experiences of the world (Thomas 1998:19-20). However, as human beings we are not merely receiving information, but previous experiences and cultural meanings of things direct what we sense and which meanings we give to the things we perceive (Thomas 1998:18). In a sense, this kind of perspective on the world means breaking down the distinctions between mind and matter and subject and object (Thomas 1998:16-20). Thus, when we interpret the relationships between humans and animals from a phenomenological or dwelling perspective, we focus on the practical involvement of humans and animals in everyday activities as well as on cultural ideas about animals, and how the involvement with animals was articulated in people's worldview (Thomas 1998:16-20; Ingold 2000:43-57; Armstrong Oma 2007:62-67).

The relationships between humans and animals in early modern Tornio were created and recreated in everyday interaction with animals, and the archaeological record can be used to trace this interaction. The archaeological animal bone material from early modern Tornio indicates a number of ways in which people interacted with animals and ways in which animals were incorporated in their everyday life. Animals were kept in the town estate, they lived in animal shelters around the courtyard and they were cared for on a daily basis. Domestic animals, especially cattle and sheep, lived long, and thus, individual animals were probably present in the household for several years. Moreover, the presence of slaughtering waste in all the plots in the town indicates that animals were probably slaughtered in the town. In seventeenth-century Tornio, different hunting activities occupied the people throughout the whole year. The townsfolk hunted various animal species and visited various environments, such as marine and freshwater coasts, rivers, swamps, open sea, marshes and forests. All these things that can be seen in the archaeological animal bone material were integrated in the everyday life of the people in early modern Tornio and were articulated in their experience of the world. They can also be considered as having been interacting or communicating with animals instead of taking actions on animals (cf. Bird-David 1999:77), and we must consider the possibility that people in Tornio also saw their interaction with animals that way.

The ideas concerning relational worldview have mainly been developed by studying hunter-gatherer societies and are generally applied only to such societies (e.g. Bird-David 1990; 1999:78; Viveiros de Castro 1998:471), and thus it may seem far-fetched to claim that people in a Western urban setting such as Tornio would have seen themselves as engaging in reciprocal relationship with multiple animal species. However, there are indications that animals may have been seen in a relational perspective in medieval and early modern Europe. For instance, the medieval and early modern trials where animals were accused of crimes or sentenced to death because of bestiality can be regarded, at least to some extent, as evidence of a conception of animals that is different from ours (Dinzelbacher 2002; Keskisarja 2006). According to Dinzelbacher (2002:420-421), some features in medieval animal trials suggest a conception of animals as capable of human-like feelings, reasoning and personality. Also, it seems that cattle and horses were sometimes considered to be more "human-like" than other livestock in early modern Finland, and human emotions and reactions were sometimes seen in cattle, whereas such conceptions were not attached to sheep, goats or hen (Keskisarja 2006:160-162). Finnish ethnographic material, collected mainly during the nineteenth and the early 20th century, also includes several examples of the ways of considering the relationships between people and animals and people and the environment in terms of interaction and mutuality. Perhaps the most striking example is the idea of forest movement. According to Tarkka (1994:59), forest and wild animals were generally seen as innocent creatures who did not meddle with people's lives. However, when they were not treated respectfully, the forest could start 'moving' which could be expressed, for instance, as a bear attacking cattle (Tarkka 1994:85-86).

There are also a number of factors that suggest the possibility of relational perception of animals in early modern northern Finland. In general, these are related to the preservation of the elements of an old, animistic-shamanistic worldview in northern Finland well into the early modern period (Herva & Ylimaunu 2009). These long-term pre-modern conceptions of the the world are evident in folk beliefs from northern Finland, and they have also been incorporated into the interpretation of archaeological finds from the area (Herva & Ylimaunu 2009). Certain archaeological finds from seventeenth-century Tornio, for instance, suggest that the relationships between people, animals and things may have been seen differently from the modern worldview. More specifically, a number of possible building or borderline deposits have been found from Keskikatu excavation areas 1 and 5, including the bear claw deposit discussed earlier, as well as ceramic vessels and an iron bar (Herva, forthcoming). According to Herva (forthcoming), these deposits give an alternative insight into how people considered the relationships between themselves and non-human things; some buildings could have special properties, which enabled some degree of personal interaction with these buildings.

In addition to the special deposits containing bear nails, pre-modern conceptions of human-animal interaction are evident in the practice of individual naming of animals. Cattle and horses were generally named individually in Finland (Lukkarinen 1939; Vuorela 1975:197) and the probate inventories from seventeenth-century and eighteenth-century Oulu include detailed descriptions of the coloring of the animals, as well as a list of cow names from 1799 (Brenner 1963-66). The probate inventories from early modern Tornio do not enlist the names of animals, but the ages and appearances of especially horses, cows and heifers are often carefully described (OMA BIa:1–9). This can be translated, at least to some extent, as a conception of the individuality and personality of these animals. However, age is also an important factor when the price of the animal has to be evaluated (Puputti 2008). The naming of cattle and horses but not other domestic animals in early modern Oulu may perhaps be taken as an indication of different attitudes towards cattle and horses than towards other domestic animals (cf. Keskisarja 2006:159-162). It is important to note, however, that individual and personal relationships with domestic animals do not inevitably mean sentimental or humanising attitudes towards them, as they were still creatures that could be turned to meat, skins and fat after they could no longer work or produce (Puputti 2008).

6.3. Attitudes towards wild animals and wilderness in seventeenth-century Tornio

This chapter discusses the attitudes people in early modern northern Finland had towards wild animals and nature, and the relevance of such categories as wild and domestic in seventeenth-century Tornio. Nature and wild animals have traditionally been regarded as something distinctively different from the domestic environment, and the dichotomy of nature and culture has been taken more or less for granted in Western scientific discourse (Descola and Pàlsson 1996). Nowadays, though, it is generally recognised that the separation of culture and nature is a distinctively Western phenomenon and cannot be uncritically applied to past societies (Descola and Pàlsson 1996; Ellen and Fukui 1996). I will question whether utilising wild and domestic resources were perceived as fundamentally different spheres of life, as it is often portrayed in ethnographic studies.

The dichotomy of wild and domestic has undoubtedly been a part of European thinking in the medieval and post-medieval periods. Hunting and wilderness were strictly ritually confined domains in medieval and post-medieval North-Western Europe (Hell 1996). Also, in Finnish-Carelian oral poetry, collected during the late nineteenth century and early twentieth century, there is a clear distinction between the farm and the forest (Anttonen 1994; Tarkka 1998). The forest and the wild animals in it are seen as something different from the social or cultural farm and village environment, and venturing to the forest was always a dangerous, liminal act (Tarkka 1998:95-100). According to Tarkka (1998:130-131), the distinction between forest and village was used in the same sense as 'us' and 'others', and the forest was seen as a reversed world. Furthermore, on one hand, wild animals, especially the bear and the wolf, were feared and despised because of the threat they posed to the cattle (Klemettinen 2002), and on the other hand, the bear was considered a powerful magical animal representing humanity in the reverse world of the forest (Tarkka 1998:96).

However, the applicability of these kinds of ideas to early modern Tornio is highly problematic. Holm (2002) argues that in Norway, the ethnographic data collected during the nineteenth century does not provide an accurate representation of the human-environmental relationships of the past. The ethnographic data is collected from an essentially agrarian community in which the forest and wild animals represent an unknown, threatening world. Yet, Holm argues, during the Iron Age the forest was continually used for hunting and cultivating remote summer farms, and was thus an everyday part of the environment. Along the same lines, one can ask what the forest and its wild animals were like to the people in seventeenth-century Tornio. Was the forest a reversed world and an 'otherness', as it is described in later oral poetry from other parts of Finland, or was it a normal part of the environment, a part of the continuum of the spheres of life?

Historical data on such issues is extremely scarce in seventeenth-century northern Finland. However, archaeological animal bone material can provide answers to this question in two ways. First, there seems to be no difference in the treatment and deposition of bones of wild and domestic animals (see Figs. 4 and 5). Contextual data of this kind is often employed in discussing attitudes towards and practices concerning certain species or groups of animals (e.g. Price 1985; Jones 1998). Nevertheless, it has to be remembered that the relationship between a set of attitudes towards an animal and the treatment of its remains is by no means simple and straightforward. Instead, the speciality of an animal may be context-dependent; it may be considered an interactive person in one situation, whereas in another, it may simply be treated as a soulless creature or merchandise (Willerslev 2001; 2007:116-118). Thus, the apparent lack of special treatment of wild animal bones does not necessarily mean that no beliefs or attitudes at all were attached to them.

There is, however, another line of evidence that suggests that the wild and the domestic may have not been valid categories in seventeenth-century Tornio. That is, the seventeenth-century animal bone assemblage indicates that the role of wild animals and hunting was exceptional in the subsistence of people in seventeenth-century Tornio. Hunting and interaction with wild animals was a regular activity, and hence, knowledge of different ecotypes, wild animal species and hunting techniques was a part of people's everyday life. Because the relationships between people and animals and people and their environment are constructed in the course of mutual interaction, the central role of wild resources in the lives of people in the seventeenth century must have had its effect on the attitudes people had towards wild animals and wild environments.

Despite the abovementioned reservations, then, the important role of wild animals in the daily subsistence in seventeenth-century Tornio and the lack of a special treatment of wild animal bones give reason to question the wild/domestic dichotomy in seventeenth-century Tornio. Together these lines of evidence suggest that wilderness was not necessarily thought of as something fundamentally different from the domestic sphere, or, at least, was probably not seen as a dangerous, reversed otherness as in the later ethnographic data. It has been suggested that among the Sami there was no such concept as nature as something unfamiliar and 'out there', but rather the environment was something that people were actively and concretely involved with (Schanche 2002:162-166). We must consider that this kind of alternative worldview may be comparable to what people in seventeenth-century Tornio also had, and that people related similarly to those wild and domestic environments that were actively used and visited.

6.4. Modernising human-animal relationships

The attitudes towards wild animals, however, changed as urbanisation and modernisation affected the economy, subsistence and worldview of the people in early modern Tornio. These changes are related to European-wide ideological currents such as Enlightenment and improvement, which brought along profound changes in the ways people saw the world and their place in it (e.g. Mrozovski 1999; Tarlow 2007). In early modern Europe, the objectification, or 'commodification' of nature and its resources and ecological imperialism are central themes in human-environmental relationships (Crosby 1986; Mrozovski 1999). According to Mrozovski (1999), nature and its resources transformed into something that could be owned, sold and improved by people. This progression is related to the ecological expansion of Europeans during this period; the Europeans set out to conquer new colonies, establish "neo-Europes" with Old World plants, animals and microbes, and exploit the ecological resources of their colonies (Crosby 1986).

Åström (1978:113) argues that the concept of ecological colonialism can also be applied to early modern northern Finland, as the area became a supplier of raw materials, such as timber and tar, to the European market. This led, at least to some degree, to a commodification of the forests in the area. Evidently, the monetary value of northern Finland's forests grew during the eighteenth century due to sawmill and tar industries and it seems probable that the condition of the forests also declined because of industrial use (Halila 1954:229–48; Massa 1994:161–8). For instance, according to a mid-nineteenth-century survey, the forests were then in poor condition throughout the coastal zone of Finland (Massa 1994:86–7). Moreover, the general parcelling out of land for farms began in northern Finland in the late eighteenth century, changing the land use pattern in the area (Halila 1954:88–92). It, for instance, determined the use of common hunting grounds, and, as the value of forests as an industrial resource grew, the state attempted to control their use by appointing supervisors (Halila 1954:232–5; Massa 1994:84–5).

These progressions are inextricably linked to human-animal relationships in early modern northern Finland. Generally speaking, the specialisation of animal goods production that relates to modernisation can deeply change the way people perceive their environment, and their interaction with animals and nature (Thomas 1984; Pálsson 1994). As discussed in Chapters 2.1. and 6.2., human-animal relationships are strongly affected by the complex interaction between cultural ideas, social dynamics and economic activities. Changes in subsistence patterns and interaction with different animal species can deeply alter the way people think about their environment or these animals (e.g.

Pálsson 1994; Klemettinen 2002; Ylimaunu 2002). As the natural resources become objectified and thought of in terms of ownership and profit, the relationship with nature can change to be more unilateral as people see nature as something that can be harvested but that does not have a meaningful relationship to them (Martin 1978; Pálsson 1994). Against this background, the narrowing subsistence base and decreasing hunting in early modern Tornio (see Fig. 8) can be seen as an indication of a change in the attitudes towards wild animals. Previously, wild animals were a part of everyday subsistence. During the eighteenth century, they may have become more of a commodity that one bought and sold alongside with other products such as timber and tar, without personal engagement with the environment. Furthermore, it can be argued that the relationship between people and wild animals became more one-sided and people perhaps started to see nature as a passive supply of raw materials.

However, this does not necessarily mean that the relationship to wild animals and nature was irreversibly altered and commodified, but the different modes of thinking about animals may have co-existed and alternated (cf. Willerslev 2001). For instance, according to Willerslev (2001) the Yukaghir hunters of Siberia can relate to game animals simply as merchandise in certain situations, whereas in other situations the animals are seen as individual beings who interact with the hunter on a personal level. Moreover, the attitudes towards different animals may alter differently. For instance, people may have related differently to wild animals that were no more a part of the daily life, while maintaining a closer relationships to the few species that were still hunted, such as arctic hares and capercaillies. And finally, the relationships between people and wild animals were by no means homogeneous in Tornio, which can be seen in the varying proportions of wild animal bones in different excavation areas (Fig. 7). The relationships between people and wild animals may have varied socially and individually. Some people were still probably actively engaging with wild animals while others may have regarded them as mere merchandise.

On a general level, the eighteenth century was a period of profound alteration in the ways people related to agriculture and animal husbandry. In Central Europe and Britain, agricultural production became more specialized and there developed a growing urban population alienated from the production of agricultural goods (Thomas 1984; Franklin 1999). At the same time, the ideologies of Enlightenment and physiocracy emphasized the value of improvements in agricultural technology and encouraged legislation relating to agricultural change (Halila 1954:181; Tarlow 2007). This led, for instance, to developments in veterinary medicine and domestic animal care, and to a growing interest in animal breeding in northern Finland (Halila 1954:213-228). Moreover, eighteenth-century historical records provide evidence for a number of changes in the urban subsistence patterns in Tornio. After the Great Northern War in the early years of the century, the number of cattle in the town decreased (Mäntylä 1971:264). At the same time, fishing activities of the townsfolk lost their former importance and the number of craftsmen in the town increased (Mäntylä 1971:248, 413). The significance of farming activities in the town probably decreased, or, at least, the area of cultivated fields diminished and the consumption of imported crops increased (Mäntylä 1971:408–9).

However, the changes that took place in the relationships between people and domestic animals in eighteenth-century Tornio seem rather minor in the light of zooarchaeological data. It seems that animals were still raised mainly for milk and wool production and there are no indications of a specialised meat cut trade (Chapter 5.2, see also Puputti 2008). Also the historical records indicates that people remained quite self-sufficient in farm products (Mäntylä 1971), agrarian characteristics remained in the urban landscape (Herva & Nurmi 2009), and domestic animals remained a part of the urban milieu in Tornio (Puputti 2008). Thus, the modernisation of the human-animal and human-environmental relationships of eighteenth-century Tornio may be seen in the narrowing down of the subsistence pattern, from a mixed subsistence pattern to an agrarian-dominated one.

The perseverance of agrarian characteristics in the urban milieu at the same time with growing alienation from wild animals and environments has interesting implications on the conceptual boundary between wild and domestic spheres of life in eighteenth-century Tornio. More specifically, the eighteenth century may very well have been the period when people began to see wild resources as something different from the domestic resources in Tornio. New archaeological data indicates that urban landscape and the way of life transformed in eighteenth-century Tornio in ways that are related to new ideas about the boundaries of culture and nature as well as public and private (Ylimaunu 2007; Herva et al. forthcoming). Archaeological investigations have revealed that the organisation of urban space in the early seventeenth century was open and village-like, while during the late seventeenth century and eighteenth century, urban space was reorganised as a new grid plan was designed and regularised and plots were enclosed (Mäntylä 1971; Ylimaunu 2007; Herva et al. forthcoming). In a way, this means that Tornio started to resemble a town in an early modern sense during the eighteenth century, with town a plan and a landscape reflecting orderliness, state's control over its people and man's control over nature (Herva et al. forthcoming; cf. Tarlow 2007). Man's control over nature is especially clear in garden cultivation, which became more common in Tornio during the eighteenth century (Mäntylä 1971:411-412). Some people, mostly wealthy merchants, established gardens in empty plots and cultivated new varieties of vegetables such as potatoes, lettuce, beetroots, parsnip and cucumber (Mäntylä 1971:412). Tarlow (2007:76-77) relates the growing interest in gardening as demarcating boundaries between domestic environment and nature and manifesting man's control over nature. According to her, "nature was controlled and ordered in the garden" (Tarlow 2007:76-77). The new gardens in Tornio, besides supplying varying vegetables to the townsfolk's diet, may also be interpreted as belonging to a new way of relating to and mastering nature.

Finally, it is important to note that human-animal relationships of a given community do not form a homogeneous entity. Rather, the relationships alter according to gender, age, social position and animal species. For instance, the interaction with certain animals may have been divided according to gender or social position. In traditional Finnish rural culture livestock husbandry, milk and wool processing, food processing, cooking and food distribution have been the duties of women (Stark-Arola 2001). On the other hand, caring for horses, hunting, fishing and working in the field and forest have been predominantly male activities (Stark-Arola 2001). According to Fjellström (1993), this kind of division of labor was, essentially, used in the Tornionjoki River Valley in the early modern period as well. Caring for animals also included age group-specific tasks; for instance herding sheep and cattle in the woods has traditionally been the task of young boys (Keskisarja 2006:124). Thus, interaction with different animal species was gender, age and social group-specific; a male head of the household had different relationships with certain animals than his wife, son or servant. If we accept that caring for livestock was an activity assigned to women and caring for horses, fieldwork and hunting were male activities, we must also conclude that the change that took place in the role of hunting was also a gendered one. The shift towards decreasing in hunting essentially took place in the everyday life and worldview of men, whereas the role of livestock husbandry remained an important part of the everyday life of women.

7. Conclusions

This study of the archaeological animal bone material from early modern Tornio revealed a number of patterns and practices of waste disposal, animal husbandry, hunting and human-animal relationships in the town.

- A pattern of the disposal and treatment of animal remains that varied according to species. Especially companion animals such as dogs, cats and horses were treated differently from other species. Although the animal bone distribution in different context types was quite homogenous, it seems that pits, ditches and heaps in the yard may have been used in the disposal of slaughtering waste.

- Livestock husbandry mainly based on cattle, sheep, pigs and reindeer, in the order of importance. There was a strong emphasis on milk and wool production, although cattle had a sex-specific culling pattern according to age. There were no indications of growing specialised meat trade during the seventeenth and eighteenth centuries.

- A major role of wild animals in the diet and subsistence of the townspeople during the seventeenth century. A large variety of species, such as Tetraonid birds, arctic hares, seals and waterfowl were hunted, although the overall emphasis was on hunting inland species.

- A possibility of reinterpreting human-animal relationships in an early modern urban environment. Human-animal relationships in seventeenth-century Tornio may be interpreted in terms of interaction and mutuality, although the attitudes towards animals probably varied according to social group and animal species.

- A clear change towards utilising a narrower resource base and domestic species during the eighteenth century, which was closely tied to all-European shift in attitudes towards nature and wild resources. At the same time, rural characteristics, such as animal husbandry, remained an important part of urban life.

The attitudes people in early modern Tornio had towards different animals, as food and living creatures, were complex and probably varied according to social position, age, gender, animal species and period. The changes that happen in the interaction between people and animals, and people and their environment must also be understood as tied to social position, gender, age and animal species, as well as to economical and ideological currents and worldview.

Appendix I. Numbers of identified specimens (NISP) in each excavation area in 1621-1660.

Common name	Scientific name	Area 1	Area 2	Area 3	Area 5	Area 6	Area 8
Cattle	*Bos taurus*	366	262	27	316	74	123
Sheep	*Ovis aries*	95	43	15	75	24	23
Goat	*Capra hircus*						4
	Ovis aries/Capra hircus	65	59	5	90	13	36
Pig	*Sus scrofa domesticus*	51	26	2	70	17	28
Reindeer	*Rangifer tarandus*	14	23	2	1	1	9
Horse	*Equus caballus*				6		1
Middle-sized ungulate	*Artiodactyla/Perissodactyla*	317	97	15	207	43	62
Large ungulate	*Artiodactyla/Perissodactyla*	454	243	18	408	73	104
Dog	*Canis familiaris*		5		1		
Canid	*Canidae*		1				
Red fox	*Vulpes vulpes*	7			1		
Cat	*Felis catus*	4					
Grey seal	*Halichoerus grypus*		1				
Ringed seal	*Phoca hispida*	2	1		1		
Seal	*Phocidae*	39	8		14	7	2
Arctic hare	*Lepus timidus*	63	46		56	7	3
Red squirrel	*Sciurus vulgaris*	2				1	
Mammal	*Mammalia*	1188	503	67	576	215	193
Black grouse	*Tetrao tetrix*	24	5		20	6	1
Capercaillie	*Tetrao urogallus*	101	34	4	75	23	15
Willow grouse	*Lagopus lagopus*	3	5		10		
Willow grouse/ Rock ptarmigan		57	4		11	6	2
Hazelhen	*Bonasa bonasia*	4			2		
Domestic chicken	*Gallus domesticus*	1		1	2		2
Gallinaceous bird	*Galliformes*	7			2		
Long-tailed duck	*Clangula hyemalis*	1					
Whooper swan	*Cygnus cygnus*	14			1	12	
Goosander	*Mergus merganser*	2			1		1
Red-breasted merganser	*Mergus serrator*	3			1		
	Mergus serrator/merganser	2			3		
Tufted duck	*Aythya fuligula*	1					
Scaup	*Aythya marila*	1					
	Aythya sp	1					
Greylag goose	*Anser anser*	1			10		2
Lesser white-fronted goose	*Anser erythropus*				3		
Bean goose	*Anser fabalis*	1	1		8		1
	Anser sp	2			7	1	2
	Anas sp	16	1		26		
Waterfowl	*Anatidae*	1	1		2		
Bird	*Aves*	161	29	3	94	5	18

Appendix I. (Continued)

Common name	Scientific name	Area 1	Area 2	Area 3	Area 5	Area 6	Area 8
Fish	*Pisces*	399	27	2	143	35	54
Undetermined		21	3		42		
	total	3491	1451	162	2285	578	702

Appendix 2. Numbers of identified specimens (NISP) in each excavation area in 1650-1728.

Common name	Scientific name	Area 1	Area 2	Area 3	Area 5	Area 6	Area 8
Cattle	Bos taurus	60	82	127	61	21	6
Sheep	Ovis aries	16	19	22	11	5	1
Goat	Capra hircus			1			
	Ovis aries/Capra hircus	13	16	2	12	6	4
Pig	Sus scrofa domesticus	4	15	22	13	4	1
Reindeer	Rangifer tarandus	4	1	4			
Middle-sized ungulate	Artiodactyla/Perissodactyla	74	81	29	40	11	13
Large ungulate	Artiodactyla/Perissodactyla	127	137	89	83	28	5
Dog	Canis familiaris			1			
Red fox	Vulpes vulpes	8					
Cat	Felis catus	1		1			
Brown bear	Ursus arctos	10					
Grey seal	Halichoerus grypus	1					
Ringed seal	Phoca hispida	1					
Seal	Phocidae	9			2		
Arctic hare	Lepus timidus	11	6	4	5	7	5
Mammal	Mammalia	182	185	179	113	54	10
Black grouse	Tetrao tetrix	4	1		5		
Capercaillie	Tetrao urogallus	3	10	2	4	9	4
Willow grouse	Lagopus lagopus	4	3	2	2		
Willow grouse/Rock ptarmigan	Lagopus sp	7	2				1
Hazelhen	Bonasa bonasia			1			
Domestic chicken	Gallus domesticus			1		1	2
Gallinaceous bird	Galliformes				1		
Whooper swan	Cygnus cygnus	2					
Scaup	Aythya marila	1					
Bean goose	Anser fabalis	1					
	Anas sp	4	2		3		2
Waterfowl	Anatidae	1		1			
Bird	Aves	20	15	2	12	4	3
Fish	Pisces	6	16	4	37	14	
Undetermined		11			3		
	total	583	603	496	407	164	60

Appendix 3. Numbers of identified specimens (NISP) in each excavation area in 1721-1800.

Common name	Scientific name	Area 2	Area 3	Area 5	Westring	Välikatu
Cattle	*Bos taurus*	22	115	34	77	36
Sheep	*Ovis aries*	1	13		1	
Sheep/goat	*Ovis aries/Capra hircus*	1	13	3	10	6
Pig	*Sus scrofa domesticus*	1	6	2	3	3
Reindeer	*Rangifer tarandus*	4	2	7	6	5
Middle-sized ungulate	*Artiodactyla/ Perissodactyla*	4	8	1	13	6
Large ungulate	*Artiodactyla/ Perissodactyla*	10	64	14	21	15
Cat	*Felis catus*					2
Grey seal	*Phoca hispida*	1				
Seal	*Phocidae*	2				
Arctic hare	*Lepus timidus*			7	1	
Mammal	*Mammalia*	62	159	73	293	115
Black grouse	*Tetrao tetrix*				1	
Capercaillie	*Tetrao urogallus*	1	2	1	10	5
Domestic chicken	*Gallus domesticus*				1	
Gallinaceous bird	*Galliformes*		1	1		
Red-breasted merganser	*Mergus serrator*			1	0	
	Anas sp				1	
Bird	*Aves*		3	1	3	
Fish	*Pisces*		3	3	8	
	total	109	391	147	449	193

Appendix 4. The skeletal frquencies of wild species as number of identified specimens (NISP), minimum number of elements (MNE) and modified anatomical units (MAU), divided by period.

1620-1660	cranium			axial skeleton			upper forelimb			upper hindlimb			extremities		
	NISP	MNE	MAU	NISP	MNE	MAU	NISP	MNE	MAU	NISP	MNE	MAU	NISP	MNE	MAU
Sciurus vulgaris							1	1	0.5	2	2	1			
Lepus timidus	12	10	5	21	1	1	66	22	11	41	19	9.5	32	7	1.75
Phocidae															
Vulpes vulpes	1	1	1							1	1	0.5	6	1	0.25
Ursus arctos													10		
Tetrao urogallus	1	1	1	23	15	15	118	34	17	66	27	13.5	47	29	7.25
Tetrao tetrix				4	4	4	31	18	9	24	7	3.5	1	1	0.25
Lagopus sp				3	3	3	61	23	11.5	33	15	7.5	2	2	0.5
Bonasa bonasia				2	2	2	2	1	0.5	2	2	1			
Anas sp	4	4	4	3	3	3	23	11	5.5	9	4	2	3	2	0.5
Anser sp.				1	1	1	32	17	8.5	2	1	0.5	4	4	1
Cygnus cygnus	4	2	2	1	1	1	19	6	3	2	2	1	1	1	0.25
Aythya sp.							2	2	1	1	1	0.5			
Clangula hyemalis										1	1	0.5			
Mergus sp.							8	4	2	4	4	2	1	1	0.25
1650-1728															
Lepus timidus	2	2	2	9	1	1	13	6	3	12	5	2.5	2	1	0.25
Phocidae															
Vulpes vulpes	1	1	1							1	1	0.5	5	1	0.25
Tetrao urogallus				4	3	3	13	4	2	8	3	1.5	12	7	2.33
Tetrao tetrix				1	1	1	6	3	1.5	5	2	1			
Lagopus sp				2	2	2	18	7	3.5				1	1	0.25
Bonasa bonasia							1	1	0.5						
Anas sp							3	1	0.5	4	3	1.5	4	3	0.75
Anser sp.										1	1	0.5			
Cygnus cygnus				1	1	1	1	1	0.5						
Aythya sp.							1	1	0.5						
1721-1800															
Lepus timidus	7	3	3	1	1	1	3	1	0.5	2	2	1	2	1	0.25
Phocidae															
Tetrao urogallus				4	3	3	12	6	3	4	3	1.5	13	7	2.33
Tetrao tetrix							3	2	1				1	1	0.25
Lagopus sp							9	4	2	2	1	0.5	1	1	0.25
Anas sp	1	1	1				1	1	0.5	1	1	0.5	1	1	0.25
Cygnus cygnus				1	1	1									
Mergus sp.							2	2	1	1	1	0.5			

Appendix 5. Age estimations based on cattle mandibulae. Age estimations based on tooth eruption and mandibular wear stages (M.W.S.) are presented.

Excavation	Date	Mandibular teeth	Age	M.W.S.
Keskikatu area 1	1620-1660	M1	over 5 months	
Keskikatu area 5	1620-1660	M1	over 5 months	
Keskikatu area 1	1620-1660	dp2-dp4, M1	5-28 months	8
Keskikatu area 1	1620-1660	dp4, M1-M2, M3 not erupted	15-24 months	23
Keskikatu area 5	1620-1660	dp4, M1-M2	15-36 months	23
Keskikatu area 1	1620-1660	dp4, M1-M2	15-36 months	23
Keskikatu area 1	1620-1660	dp4, M1, M2, M3 not erupted	15-24 months	25-26
Keskikatu area 5	1620-1660	P3, dp4, M1-M3	24-36 months	28
Keskikatu area 5	1620-1660	dp4, M1-M3	24-36 months	28
Keskikatu area 5	1620-1660	dp3-dp4, M1-M3	24-30 months	29
Keskikatu area 5	1620-1660	dp3-dp4, M1-M3	24-30 months	29
Keskikatu area 1	1620-1660	M1-M3	over 24 months	33-35
Keskikatu area 1	1620-1660	P4-M3	over 28 months	36-37
Keskikatu area 1	1620-1660	P4-M3	over 28 months	38
Keskikatu area 1	1620-1660	P2-M3	over 28 months	40
Keskikatu area 1	1620-1660	M1-M3	over 24 months	41
Keskikatu area 1	1620-1660	P4-M3	over 28 months	42
Keskikatu area 5	1620-1660	P2-M3	over 28 months	44
Keskikatu area 1	1620-1660	P4-M3	over 28 months	49
Keskikatu area 8	1620-1660	P4-M3	over 28 months	43
Keskikatu area 8	1620-1660	dp4, M1, M2	15-36 months	25-26
Keskikatu area 8	1620-1660	M1-M3	over 24 months	36
Keskikatu area 8	1620-1660	P4-M3	over 28 months	40
Keskikatu area 8	1620-1660	P2-4	over 28 months	
Keskikatu area 4	1620-1660	P2-3, dp4, M1, M2, M3 not erupted	15-36 months	19-20
Keskikatu area 6	1620-1660	P3 erupting	under 30 months	
Keskikatu area 2	1620-1660	P3-M3	over 28 months	36
Keskikatu area 2	1620-1660	dp3-dp4	under 30 months	
Keskikatu area 2	1620-1660	dp4	under 30 months	
Keskikatu area 5	1650-1728	dp3-dp4, M1 not erupted	under 6 months	
Keskikatu area 5	1650-1728	P3, dp4, M1-M2	18-36 months	23
Keskikatu area 6	1650-1728	I erupting	14-40 months	
Keskikatu area 2	1650-1728	M2	over 24 months	
Keskikatu area 2	1650-1728	dp2-4, M1 not erupted	under 30 months	
Keskikatu area 2	1650-1728	P2-P4, M1 and M2 broken	over 28 months	
Keskikatu area 3	1721-1800	P4-M3	over 28 months	35
Keskikatu area 3	1721-1800	P2-M3	over 28 months	35
Keskikatu area 3	1721-1800	P2-3, P4 erupting, M1-2, M3 erupting	24-36 months	31
Keskikatu area 2	1721-1800	P4 erupting, M1-3	over 28 months	33
Westring	1721-1800	P2-M1	over 28 months	42
Keskikatu area 5	17th century	dp2-dp4, M1 not erupted	under 6 months	
Keskikatu area 5	17th century	dp2-dp4, M1 not erupted	under 6 months	
Keskikatu area 5	17th century	dp2-dp4, M1	5-36 months	8
Keskikatu area 5	17th century	dp2-dp4, M1	5-36 months	8
Keskikatu area 5	17th century	dp2-dp4, M1-M2, M3 not erupted	15-24 months	8
Keskikatu area 5	17th century	P4, M2	over 28 months	
Keskikatu area 5	17th century	P3-P4	over 28 months	
Keskikatu area 5	17th century	P2-P3, M1-M3, P4 not erupted	30-36 months	31
Keskikatu area 5	17th century	P3-M3	over 28 months	42

Appendix 5. (Continued)

Excavation	Date	Mandibular teeth	Age	M.W.S.
Keskikatu area 5	17th century	P4-M1	over 28 months	50
Keskikatu area 5	17th century	P2-M3	over 28 months	42
Keskikatu area 5	17th century	P4-M1	over 28 months	50
Keskikatu area 6	17th century	M1-3	over 24 months	29
Keskikatu area 6	17th century	P2, P3 not erupted, P4, M1, M2	28-30 months	39-45
Keskikatu area 6	17th century	M1-M3	over 24 months	45
Keskikatu area 6	17th century	dp3-dp4	under 38 months	
Keskikatu area 6	17th century	dp4	under 36 months	
Keskikatu area 6	17th century	P3, broken P4, M1-2	over 28 months	
Keskikatu area 6	17th century	P3-M3	over 28 months	33
Keskikatu area 6	17th century	P2-M3	over 28 months	49
Keskikatu area 2	17th century	dp2-dp4	under 30 months	
Aspio&Viippola	17th century	P3 erupting, dp4, M1-3	24-36 months	29
Aspio&Viippola	17th century	P3-4, M1-3	over 28 months	36
Aspio&Viippola	17th century	P3-4, M1-3	over 28 months	36
Aspio&Viippola	17th century	M3	over 24 months	
Aspio&Viippola	17th century	P4	over 28 months	
Aspio&Viippola	17th century	dp4	under 36 months	
Aspio&Viippola	17th century	P2, P3, P4	over 28 months	

Appendix 6. Sex assessments of cattle based on the pelvis. The minimum thickness of the medial wall of the acetabulum and the morphology of the fossa musculus rectus femoris and pubic bone are presented.

Excavation	Date	Acetab. medial thickness (mm)	Fossa musculus rectus femoris	Pubis	Sex
Keskikatu excavation area 1	17th century	4.5	f		f
Keskikatu excavation area 1	17th century		m		m
Keskikatu excavation area 1	17th century		m		m
Keskikatu excavation area 1	17th century	10	f		f
Keskikatu excavation area 1	1620-1660		f		f
Keskikatu excavation area 1	17th century			f	f
Keskikatu excavation area 1	1620-1660		f		f
Keskikatu excavation area 1	1620-1660	10.5	m		m
Keskikatu excavation area 1	1620-1660		f	f	f
Keskikatu excavation area 2	1620-1660	8.4	f		f
Keskikatu excavation area 2	1620-1660	5.29	f		f
Keskikatu excavation area 2	1620-1660	10.5	m		m
Keskikatu excavation area 2	1620-1660	14.24	m		m
Keskikatu excavation area 2	1620-1660	6.73		f	f
Keskikatu excavation area 2	1620-1660	6.96	f		f
Keskikatu excavation area 2	1620-1660	7.54			f
Keskikatu excavation area 2	1620-1660	7.4	f		f
Keskikatu excavation area 2	1620-1660		m		m
Keskikatu excavation area 2	1620-1660			f	f
Keskikatu excavation area 2	1620-1660		f		f
Keskikatu excavation area 2	1620-1660		f		f
Keskikatu excavation area 2	1620-1660		m		m
Keskikatu excavation area 2	1620-1660		f		f
Keskikatu excavation area 2	1620-1660			f	f
Keskikatu excavation area 2	1721-1800		f		f
Keskikatu excavation area 3	1620-1660	5.43	f	f	f
Keskikatu excavation area 3	1620-1660	7.09	f		f
Keskikatu excavation area 3	1721-1800		m		m
Keskikatu excavation area 3	1620-1660		m		m
Keskikatu excavation area 3	1650-1728		f	f	f
Keskikatu excavation area 3	1721-1800			f	f
Keskikatu excavation area 5	1650-1728	9	f		f
Keskikatu excavation area 5	1650-1728			m	m
Keskikatu excavation area 5	1650-1728			f	f
Keskikatu excavation area 5	17th century	8			f
Keskikatu excavation area 5	17th century			m	m
Keskikatu excavation area 5	17th century			f	f
Keskikatu excavation area 5	17th century		f		f
Keskikatu excavation area 5	17th century	10.5		m	m
Keskikatu excavation area 5	17th century			f	f
Keskikatu excavation area 5	1650-1728			m	m
Keskikatu excavation area 5	1650-1728	6.5	f		f
Keskikatu excavation area 5	1620-1660	14.4	m		m
Keskikatu excavation area 5	1620-1660		m		m
Keskikatu excavation area 5	1620-1660		f	f	f
Keskikatu excavation area 5	1620-1660	7	f		f
Keskikatu excavation area 6	17th century	14.34		m	m
Keskikatu excavation area 6	17th century			f	f

Appendix 6. (Continued)

Excavation	Date	Acetab. medial thickness (mm)	*Fossa musculus rectus femoris*	Pubis	Sex
Keskikatu excavation area 6	17th century		m		m
Keskikatu excavation area 6	17th century		m		m
Keskikatu excavation area 6	17th century			f	f
Keskikatu excavation area 6	17th century		f		f
Keskikatu excavation area 6	17th century		f		f
Keskikatu excavation area 6	17th century			f	f
Keskikatu excavation area 6	17th century		m		m
Keskikatu excavation area 6	17th century		m		m
Keskikatu excavation area 6	17th century			f	f
Keskikatu excavation area 6	17th century	8.31	f		f
Keskikatu excavation area 6	17th century	7.13	f	f	f
Keskikatu excavation area 6	17th century		m		m
Keskikatu excavation area 6	17th century			f	f
Keskikatu excavation area 6	17th century			f	f
Keskikatu excavation area 6	17th century		m		m
Keskikatu excavation area 8	1620-1660	6.37			f
Keskikatu excavation area 8	1620-1660	8.53	f		f
Keskikatu excavation area 8	1620-1660		m		m
Keskikatu excavation area 8	1620-1660			f	f
Keskikatu excavation area 8	17th century	8.94	f		f
Keskikatu excavation area 8	17th century	8.7	f		f
Keskikatu excavation area 8	17th century		f	f	f
Aspio & Viippola	17th century		m		m
Aspio & Viippola	17th century		m		m
Aspio & Viippola	17th century			f	f
Aspio & Viippola	17th century		f		f
Purra & Aho	17th century		f		f
Purra & Aho	17th century		f		f
Purra & Aho	17th century			f	f
Purra & Aho	17th century			f	f
Westring	1721-1800			f	f

Appendix 7. The measurements (mm) of cattle metacarpals and the slenderness index (100*(mc Bd/mc GL)).

Excavation	Date	mc GL	mcSD	mc Bp	mc depth p	mc Bd	mc depth d	Slenderness index
Keskikatu Area 1	1620-1660			53,81	32,29			
Keskikatu Area 2	1650-1728	173						
Keskikatu Area 2	1650-1728	174,5	22,57	43,36	28,39	45,79	25,49	26,24
Keskikatu Area 2	17th century	170,5	24,4			47,12	27,3	27,64
Keskikatu Area 2	1620-1660	173	20,42					
Keskikatu Area 2	1620-1660	173	22,94					
Keskikatu Area 3	1650-1728	185,5						
Keskikatu Area 4	1620-1660	160,5	25,9	45,58	27,61	48,07	24,63	29,95
Keskikatu Area 5	1650-1728	157	26,38	50,11	29,12	49,28	25,49	31,39
Keskikatu Area 5	17th century	154	26,41	51,33	29,03	50,29	26,36	32,66
Keskikatu Area 5	1620-1660	170						
Keskikatu Area 6	1620-1660	160	23,84					
Keskikatu Area 6	1620-1660	164,5	23,57					
Keskikatu Area 8	1620-1660	169	24,12	47,3	26,79	45,09	25,63	26,68
Keskikatu Area 8	1620-1660	177	24,93	49,12	29,9	48,79	27,94	27,56
Keskikatu Area 8	1620-1660	168,5	25,71	45,34	27,1	45,66	23,57	27,10

Appendix 8. The mean, number of specimens, standard deviation and range of postcranial measurements (mm) of cattle, ovicaprids and pigs.

	Bos taurus	Ovis aries/Capra hircus	Sus scrofa	Rangifer tarandus
hum GL		131.8		
		1		
hum BT	59.7	23.9	24.9	40.8
	17	33	15	15
	±4.4	±1.5	±1.1	±4.0
	54.2 - 69.2	20.2 - 27.4	22.2 - 26.5	31.3 - 46.8
hum Bd	63	25.5	29	44.9
	19	36	13	15
	±3.9	±1.5	±1.2	±5.3
	57.0 - 70.9	23.0 - 29.6	26.3 - 30.7	32.1 - 51.4
hum trochlear height	35.1	15.8	22.4	32.4
	20	35	14	18
	±3.1	±1.2	±1.8	±2.6
	30.6 – 42.0	13.4 - 18.5	17.7 - 24.8	28.4 38.7
hum Bp	71.67	33.6		61
	1	6		1
		±2.4		
		31.1 - 37.4		
hum HHAP	54.3	24.5		42.7
	2	5		2
	±12.1	±1.6		±1.9
	45.7 - 62.9	22.7 - 26.7		41.4 - 44.0
rad GL		133.4		
		5		
		10.3		
		120.3 - 148.6		
rad PRAP	30.4	12.4	14.7	23.2
	22	34	5	2
	±2.5	±1.1	±0.7	±3.0
	25.0 - 36.3	10.8 - 14.5	13.6 - 15.6	22.4 - 24.1
rad Bp	66.7	26.9	24.2	42.7
	19	34	5	2
	±4.6	±1.6	±1.0	±3.0
	59.8 – 77.0	23.1 - 29.8	23.5 - 26.0	40.6 - 44.8
rad Bd	57.3	25.4		38.9
	12	19		1
	±3.7	±1.3		
	54.1 - 67.8	23.1 - 28.9		
fem GL		181.8		
		2		
		±15.9		
		170.5 - 193		
fem Bp	97.8	40.5		67.1
	1	15		3
		±4.6		±4.0
		34.9 - 48.3		62.5 - 69.8

Appendix 8. (Continued)

	Bos taurus	Ovis aries/Capra hircus	Sus scrofa	Rangifer tarandus
fem DG	36.4	18.4		27.3
	11	19		5
	±2.0	±1.7		±1.5
	33.6 – 41.0	16.1 - 22.4		26.3 - 30.0
fem Bd	73.9	32.6		57
	2	11		5
	±4.4	±1.9		±7.0
	70.8 – 77.0	30.4 - 35.4		45.7 - 62.4
tib GL	284	179.5		
	1	3		
		4.8		
		174.0 – 183.0		
tib Bp	78.1	35.9		
	3	10		
	±10.0	±1.1		
	67.9 – 88.0	34.1 – 38.0		
tib PTAP	77	31.8		
	1	10		
		±2.3		
		28.1 - 36.1		
tib Bd	50.3	22.5	24.9	43.7
	13	56	4	1
	±2.4	±1.1	±1.9	
	47.7 - 55.2	20.0 - 25.7	22.6 - 27.2	
mc GL	170.5	111.1		
	20	18		
	±8.1	±8.7		
	154.0 - 185.5	94.4 - 133.6		
mcSD	24.8	12		
	17	17		
	±2.1	±1.2		
	20.4 - 29.4	10.3 - 14.7		
mc Bp	48.5	20		
	13	20		
	±3.4	±1.5		
	43.4 – 54.0	18.0 - 22.7		
mc depth p	29	14.3		
	13	20		
	±2.2	±0.9		
	26.1 - 33.6	13.2 - 15.9		
mc Bd	48.3	22		
	13	18		
	±2.7	±1.8		
	45.1 - 54.3	19.8 - 24.8		
mc depth d	26.6	13.9		
	13	18		
	±1.6	±1.0		
	23.6 - 28.9	12.5 - 15.8		

Appendix 8. (Continued)

	Bos taurus	Ovis aries/Capra hircus	Sus scrofa	Rangifer tarandus
mt GL	194	118.9		
	14	8		
	±6.2	±14.5		
	185.5 – 204.0	94.5 - 142.3		
mtSD	21.7	10.9		
	13	7		
	±1.1	±2.1		
	20.1 - 24.1	8.9 - 15.0		
mt Bp	39.8	18.7		
	12	10		
	±3.6	±2.0		
	35.3 - 48.7	16.6 - 22.9		
mt depth p	38.6	17.8		
	10	9		
	±2.0	±1.3		
	35.4 - 41.7	16.5 - 19.7		
mt Bd	43.5	21.3		
	10	8		
	±6.9	±1.8		
	24.8 - 50.6	19.6 - 24.4		
mt depth d	25.5	14.1		23.1
	9	8		1
	±1.3	±1.3		
	22.9 - 26.9	12.9 - 15.9		

Appendix 9. The log ratios of postcranial skeletal measurements (means, sample sizes and standard deviations) of cattle, ovicaprids, pig and reindeer divided by period.

	Bos taurus	Ovis aries/Capra hircus	Sus scrofa	Rangifer tarandus
	0.012	0.002	-0.044	0.021
	124	253	29	34
1620-1660	±0.03	±0.04	±0.02	±0.05
	-0.002	-0.001	-0.001	0.010
	48	51	3	3
1650-1728	±0.02	±0.05	±0.02	±0.01
	-0.006	-0.012	0.009	0.018
	43	18	2	6
1721-1800	±0.03	±0.03	±0.04	±0.03

Appendix 10. The skeletal frequencies of cattle as number of identified specimens (NISP), minimum number of elements (MNE) and modified anatomical units (MAU), divided by period.

	1621-1660			1650-1728			1721-1800		
	NISP	MNE	MAU	NISP	MNE	MAU	NISP	MNE	MAU
cranium (mandibula)	329	16	16	86	4	4	47	8	8
cervical vertebrae (atlas)	33	9	9	16	5	5	7	3	3
thoracic vertebrae	23	2	2	11	1	1	3	1	1
lumbar vertebrae	15	3	3	5	1	1	5	1	1
sacrum	3	3	3	2	2	2	1	1	1
caudal vertebrae	9	1	1	2	1	1			
scapula (prox)	75	37	18.5	14	9	4.5	12	5	2.5
humerus (dist)	73	25	12.5	16	9	4.5	7	2	1
radius (prox)	48	31	15.5	13	8	4	10	6	3
ulna (prox)	33	31	15.5	9	8	4	6	6	3
carpal bones (ci)	44	12	6	17	3	1.5	5	2	1
metacarpal bone (prox)	19	16	8	8	7	3.5	2	1	0.5
phalanges (phalanx proximalis)	159	8	2	65	4	1	37	3	0.75
pelvis (acetabulum)	64	26	13	23	7	3.5	9	4	2
femur (caput femoris)	70	24	12	12	4	2	9	6	3
tibia (dist)	27	20	10	10	8	4	2	2	1
patella	15	15	7.5	3	3	1.5	2	2	1
malleolare	3	3	1.5	2	2	1			
tarsal bones (talus)	55	26	13	19	9	4.5	9	3	1.5
metatarsal bone (prox)	22	22	11	24	12	11	1	1	0.5

Appendix 11. Age estimations based on ovicaprid mandibulae. Age estimations based on tooth eruption and mandibular wear stages (M.W.S.) are presented.

Excavation	Date	Mandibular teeth	Age	M.W.S.
Keskikatu excavation area 5	1620-1660	dp2-dp4, M1-M2	8-30 months	21
Keskikatu excavation area 8	1620-1660	P2-4 erupting, M1-M2	8-30 months	22
Keskikatu excavation area 8	1620-1660	dp4, M1-2, M3 not erupted	8-30 months	30
Keskikatu excavation area 8	1620-1660	dp4, M1-2, M3 not erupted	8-30 months	21
Keskikatu excavation area 2	1620-1660	P2-3, dp4, M1-2, M3 not erupted	8-30 months	24
Keskikatu excavation area 2	1620-1660	dp4, M1-2, M3 not erupted	8-30 months	25
Keskikatu excavation area 5	1620-1660	P2-P4, M1-M2, M3 not erupted	17-30 months	27
Keskikatu excavation area 2	1620-1660	P3-M3	over 18 months	31
Keskikatu excavation area 6	1620-1660	P3-M3	over 18 months	34
Keskikatu excavation area 2	1620-1660	P2-M1	over 17 months	28-35
Keskikatu excavation area 2	1620-1660	P3-M3	over 18 months	40
Keskikatu excavation area 2	1620-1660	M2-M3	over 18 months	45-49
Keskikatu excavation area 2	1620-1660	M2-M3	over 18 months	39
Keskikatu excavation area 2	1620-1660	P2	over 17 months	
Keskikatu excavation area 2	1620-1660	M3	over 18 months	
Keskikatu excavation area 2	1620-1660	P2-P3	over 17 months	
Keskikatu excavation area 5	1620-1660	P2-M2	over 17 months	35-37
Keskikatu excavation area 5	1620-1660	P2-M3	over 18 months	38
Keskikatu excavation area 5	1620-1660	P3-P4, M1-M3	over 18 months	41
Keskikatu excavation area 8	1620-1660	P2-M3	over 18 months	42
Keskikatu excavation area 1	1650-1728	dp4, M1-M3 (M3 erupting)	18-30 months	24
Keskikatu excavation area 5	1650-1728	P4-M3	over 18 months	31
Keskikatu excavation area 5	1721-1800	P4-M3	over 18 months	37
Keskikatu excavation area 3	17th century	P2-3, dp4, M1, M2 erupting	5-12 months	11
Keskikatu excavation area 6	17th century	P3, dp4, M1-2, M3 not erupted	8-30 months	22-23
Keskikatu excavation area 6	17th century	P2-3 erupting, dp4, M1-2, M3 not erupted	17-30 months	20
Aspio & Viippola	17th century	P2-M3	over 18 months	34
Keskikatu excavation area 1	17th century	P4-M3	over 18 months	31
Keskikatu excavation area 6	17th century	P2-4, M1-2	over 17 months	30-33
Keskikatu excavation area 6	17th century	P2-M3	over 18 months	31
Keskikatu excavation area 6	17th century	P3-M3	over 18 months	35
Keskikatu excavation area 6	17th century	P3-M3	over 18 months	35
Keskikatu excavation area 6	17th century	M3	over 18 months	
Keskikatu excavation area 2	17th century	P3-M3	over 18 months	35
Keskikatu excavation area 2	17th century	P3-M3	over 18 months	30
Aho&Purra	17th century	P3-M3	over 18 months	41
Keskikatu excavation area 1	17th century	P4-M3	over 18 months	39
Keskikatu excavation area 8	17th century	P3-P4, M1-M3	over 18 months	41

Appendix 12. Sex assessments of sheep based on the morphology of the pelvis.

Excavation	Date	Sex
Keskikatu excavation area 1	1620-1660	m
Keskikatu excavation area 1	1650-1728	f
Keskikatu excavation area 1	1650-1728	f
Keskikatu excavation area 1	17th century	m
Keskikatu excavation area 1	1620-1660	f
Keskikatu excavation area 1	1620-1660	f
Keskikatu excavation area 1	17th century	m
Keskikatu excavation area 1	1620-1660	f
Keskikatu excavation area 1	17th century	f
Keskikatu excavation area 1	17th century	m
Keskikatu excavation area 2	1650-1728	f
Keskikatu excavation area 2	17th century	m
Keskikatu excavation area 2	1620-1660	m
Keskikatu excavation area 2	1620-1660	m
Keskikatu excavation area 2	1620-1660	f
Keskikatu excavation area 2	1620-1660	f
Keskikatu excavation area 3	17th century	f
Keskikatu excavation area 3	17th century	f
Keskikatu excavation area 3	17th century	f
Keskikatu excavation area 3	1721-1800	f
Keskikatu excavation area 3	1650-1728	m
Keskikatu excavation area 3	1620-1660	f
Keskikatu excavation area 4	1650-1728	m
Keskikatu excavation area 4	1620-1660	m
Keskikatu excavation area 4	1620-1660	f
Keskikatu excavation area 5	17th century	f
Keskikatu excavation area 5	17th century	m
Keskikatu excavation area 5	17th century	m
Keskikatu excavation area 5	17th century	f
Keskikatu excavation area 5	1620-1660	m
Keskikatu excavation area 5	1620-1660	f
Keskikatu excavation area 5	1620-1660	f
Keskikatu excavation area 5	1620-1660	f
Keskikatu excavation area 5	1650-1728	f
Keskikatu excavation area 5	1620-1660	m
Keskikatu excavation area 5	1620-1660	f
Keskikatu excavation area 5	1620-1660	m
Keskikatu excavation area 6	17th century	m
Keskikatu excavation area 6	17th century	m
Keskikatu excavation area 6	17th century	f
Keskikatu excavation area 6	17th century	m
Keskikatu excavation area 6	17th century	m
Keskikatu excavation area 6	17th century	m
Keskikatu excavation area 6	17th century	m
Keskikatu excavation area 6	17th century	m
Keskikatu excavation area 6	1620-1660	m
Keskikatu excavation area 6	1620-1660	f
Keskikatu excavation area 6	1650-1728	m
Keskikatu excavation area 6	1620-1660	m
Keskikatu excavation area 6	1620-1660	m
Keskikatu excavation area 8	1620-1660	f

Appendix 12. (Continued)

Excavation	Date	Sex
Keskikatu excavation area 8	1620-1660	m
Purra & Aho	17th century	m
Purra & Aho	17th century	m
Aspio & Viippola	17th century	f
Aspio & Viippola	17th century	m
Aspio & Viippola	17th century	f

Appendix 13. The skeletal frequencies of sheep or goat as number of identified specimens (NISP), minimum number of elements (MNE) and modified anatomical units (MAU), divided by period.

	1621-1660			1650-1728			1721-1800		
	NISP	MNE	MAU	NISP	MNE	MAU	NISP	MNE	MAU
cranium	82	13	13	15	3	3	11	3	3
cervical vertebrae	18	3	3	7	1	1	1	1	1
thoracic vertebrae	23	2	2	2	1	1	1	1	1
lumbar vertebrae	17	3	3	8	2	2			
sacrum	1	1	1						
scapula	30	15	7.5	9	3	1.5	2	2	1
humerus	46	31	15.5	6	3	1.5	1	1	0.5
radius	36	22	11	16	8	4	3	2	1
ulna	21	21	10.5	3	2	1	3	3	1.5
carpal bones	3	2	1	2	2	1			
metacarpal bone	19	19	9.5	3	3	1.5			
phalanges	43	3	0.75	14	2	0.5	1	1	0.25
pelvis	43	20	10	10	4	2	3	1	1
femur	56	24	12	8	4	2	2	2	1
tibia	54	38	19	8	6	3	4	2	1
patella	2	2	1	1	1	0.5			
tarsal bones	31	16	8	14	5	2.5			
metatarsal bone	16	16	8	4	4	2			

Appendix 14. Age estimations based on pig mandibulae. Age estimations based on tooth eruption are presented.

Excavation	Date	Mandibular teeth	Age
Keskikatu excavation area 1	1620-1660	I	over 17 months
Keskikatu excavation area 1	1620-1660	P2-M3, M3 erupting	17-22 months
Keskikatu excavation area 1	1620-1660	P2-P4	over 12 months
Keskikatu excavation area 1	1620-1660	P3-M3	over 17 months
Keskikatu excavation area 1	1620-1660	P4-M2	12-17 months
Keskikatu excavation area 2	1620-1660	M2-3	over 17 months
Keskikatu excavation area 2	1620-1660	dp2-dp4, M1 erupting	4-6 months
Keskikatu excavation area 2	1620-1660	I, C, dp2, P3 erupting, P4, M1	12-17 months
Keskikatu excavation area 2	1620-1660	M3	over 17 months
Keskikatu excavation area 2	1620-1660	dp3-dp4, M1. M2-M3 not erupted	4-7 months
Keskikatu excavation area 2	1620-1660	P3-P4, M1-M2, M3 not erupted	12-22 months
Keskikatu excavation area 2	1620-1660	dp2-dp4, M1, M2 not erupted	4-7 months
Keskikatu excavation area 5	1620-1660	P2-M2, M3 not erupted	12-22 months
Keskikatu excavation area 5	1620-1660	P3-M2, M3 not erupted	12-22 months
Keskikatu excavation area 5	1620-1660	dp2-dp4	under 16 months
Keskikatu excavation area 5	1620-1660	P2-M3	over 17 months
Keskikatu excavation area 8	1620-1660	C, P2-P4	over 12 months
Keskikatu excavation area 8	1620-1660	P3-4, M1-2, M3 not erupted	12-22 months
Keskikatu excavation area 8	1620-1660	dp3-4, M1 erupting	4-6 months
Keskikatu excavation area 3	1721-1800	M2	over 7 months
Keskikatu excavation area 3	1721-1800	M1-M2, M3 erupting	17-22 months
Keskikatu excavation area 3	1721-1800	P2	over 12 months
Westring	1721-1800	I, P3, dp4, M1 (M2 not erupted)	12 months
Välikatu	1721-1800	dp4	under 12 months
Keskikatu excavation area 5	17th century	P3 ja P4, M1, M2	over 17 months
Keskikatu excavation area 5	17th century	P2-P4, M1-M3	over 17 months
Keskikatu excavation area 6	17th century	P2-M3	over 17 months
Keskikatu excavation area 6	17th century	P4-M2, M3 not erupted	12-22 months
Keskikatu excavation area 6	17th century	P2-M2	12-22 months
Keskikatu excavation area 6	17th century	P3 erupting, P4, M1-M2 (M3 not erupted)	12-17 months
Keskikatu excavation area 6	17th century	dp2-4, M1 erupting	4-6 months
Keskikatu excavation area 8	17th century	P4-M3	over 17 months
Keskikatu excavation area 8	17th century	P3-4, M1-2, M3 not erupted	12-22 months
Keskikatu excavation area 8	17th century	C, P2	over 12 months

Appendix 15. Sex assessments of pig based on the morphology of the canine.

Excavation	Date	Sex
Keskikatu excavation area 1	1621-1660	m
Keskikatu excavation area 1	1621-1660	m
Keskikatu excavation area 1	1621-1660	f
Keskikatu excavation area 1	1621-1660	m
Keskikatu excavation area 1	1621-1660	m
Keskikatu excavation area 2	1650-1721	f
Keskikatu excavation area 3	1621-1660	f
Keskikatu excavation area 5	17th century	f
Keskikatu excavation area 5	1650-1721	f
Keskikatu excavation area 5	1650-1721	f
Keskikatu excavation area 5	1621-1660	m
Keskikatu excavation area 5	1621-1660	m
Keskikatu excavation area 6	1621-1660	f
Keskikatu excavation area 6	17th century	m
Keskikatu excavation area 8	17th century	m
Keskikatu excavation area 8	1621-1660	m

Appendix 16. The skeletal frquencies of pig as number of identified specimens (NISP), minimum number of elements (MNE) and modified anatomical units (MAU), divided by period.

	1621-1660			1650-1728			1721-1800		
	NISP	MNE	MAU	NISP	MNE	MAU	NISP	MNE	MAU
cranium	62	16	8	19	14	7	3	2	1
cervical vertebrae	7	1	1						
thoracic vertebrae	4	1	1	1	1	1	2	1	1
lumbar vertebrae	2	1	1						
sacrum				1	1	1			
scapula	8	3	1.5	1	1	0.5	1	1	0.5
humerus	12	10	5	7	5	2.5			
radius	8	6	3	3	3	1.5	1	1	0.5
ulna	7	7	3.5	4	2	1			
carpal bones	15	5	2.5	6	2	1			
metacarpal bone	4	1	0.5	3	2	1			
phalanges	11	1	0.25	2	1	0.25	1	1	0.25
pelvis	11	11	5.5	2	1	0.5			
femur	19	6	3	5	4	2			
tibia	8	8	4	1	1	0.5	1	1	0.5
fibula	1	1	0.5	3	3	1.5			
patella	1	1	0.5						
tarsal bones	12	6	3	2	1	0.5			
metatarsal bone	3	1	0.5	2	1	0.5			

Appendix 17. The skeletal frequencies of reindeer as number of identified specimens (NISP), minimum number of elements (MNE) and modified anatomical units (MAU), divided by period.

	1621-1660			1650-1728			1721-1800		
	NISP	MNE	MAU	NISP	MNE	MAU	NISP	MNE	MAU
cranium	5	2	1				2		
scapula	4	1	0.5	2	2	1	2	2	1
humerus	14	13	6.5	1	1	0.5	1	1	0.5
radius	1	1	0.5						
ulna	1	1	0.5						
carpal bones	2	1	0.5				2	1	0.5
phalanges	11	1	0.25	2	1	0.25	1	1	0.25
pelvis	2	2	1				1	1	0.5
femur	3	2	1	4	3	1.5	4	2	1
tibia	1	1	0.5						
malleolare	1	1	0.5						
tarsal bones							1	1	0.5
metatarsal bone	4	3	1.5						

References

Alamäki, Yrjö 1956. *Pohjois-Pohjanmaan maataloudesta 1600-luvun alkupuolella*. Unpublished laudatur thesis, University of Helsinki, Helsinki.

Albarella, Umberto 2002. 'Size matters': how and why biometry is still important in zooarchaeology. Dobney, Keith and O'Connor, Terry (eds). *Bones and the man. Studies in honour of Don Brothwell*. Oxbow Books, Oxford.

Anttonen, Veikko 1994. Erä- ja metsäluonnon pyhyys. *Metsä ja metsänviljaa*. Kalevalaseuran vuosikirja 73, pp. 24-35.

Apo, Satu 2001. *Viinan voima. Näkökulmia suomalaisten kansanomaiseen alkoholiajatteluun ja –kulttuuriin*. Suomalaisen Kirjallisuuden Seura, Helsinki.

Armitage, Philip L. 1982. Studies on the remains of domestic livestock from Roman, medieval and early modern London: objectives and methods. In Hall, A. R. &. Kenward, H. K (eds.). *Environmental Archaeology in the Urban Context*. The Council for British Archaeology, London 1982, p. 94-106

Armstrong Oma, Kristin 2007. *Human-Animal Relationships. Mutual Becomings in Scandinavian and Sicilian Households 900-500 BC*. Unipub, Oslo.

Barone, Robert 1999. *Anatomié comparée des mammifères domestiques. Tome 1. Ostéologie*. Vigot Frères, Paris.

Bartosiewicz, László 1995. *Animals in the Urban Landscape in the Wake of the Middle Ages. A Case Study from Vác, Hungary*. BAR International Series 609. BAR Publishing, Oxford.

Bartosiewicz, László 2003. Urban landscapes and animals. Lászlovsky, József & Szabó, Péter (eds.). *People and Nature in Historical Perspective*. CEU Department of Medieval Studies & Archaeolingua, Budapest. pp. 107-120.

Berlin, Herved 1932. *De svenska nötboskaprasernas härstamning*. C.W.K. Gleerups Förlag, Lund.

Berteaux, D. and Guintard, C. 1995. Osteometric study of the metapodials of Amsterdam Island feral cattle. *Acta Theriologica* 40(1), 97–110.

Bird-David, Nurit 1990. The giving environment: Another perspective to the economic system of gatherer-hunters. *Current Anthropology* 31:2, 189-196.

Bird-David, Nurit 1999. "Animism" revisited. Personhood, environment, and relational epistemology. *Current Anthropology* 40:67-91.

Boessneck, Joachim 1969. Osteological differences between sheep (Ovis aries Linné) and goat (Capra hircus Linné). In D. Brothwell & E. Higgs (eds). *Science in Archaeology. A Survey of Progress and Research*. 2nd edition. Thames and Hudson, London, 331-358.

Bowen, Joanne 1998. To market, to market : animal husbandry in New England. *Historical Archaeology* 32(3): 137-152.

Brenner, Alf 1963-66. *Oulun kaupungin perunkirjoituksia 1653-1800*. Suomen sukututkimusseuran julkaisuja XXV I-III. 1963-66.

Brightman, Robert A. 1993. *Grateful Prey. Rock Cree Human-Animal Relationships*. University of California Press, Berkeley.

Brück, Joanna 1999: Ritual and rationality: some problems of interpretation in European archaeology. *European Journal of Archaeology* 2:313-344.

Bull, Gail & Payne, Sebastian 1982. Tooth eruption and epiphysial fusion in pigs and wild boar. In Wilson, Bob, Grigson, Caroline & Payne, Sebastian (eds). *Ageing and Sexing Animal Bones from Archaeological Sites*. BAR British Series 109, 55-72.

Bullock, D. & Rackham, J. 1982. Epiphysial fusion and tooth eruption of feral goats from Moffatdale, Dumfries and Galloway, Scotland. In Wilson, Bob, Grigson, Caroline & Payne, Sebastian (eds). *Ageing and Sexing Animal Bones from Archaeological Sites*. BAR British Series 109, 73-80.

Cannon, Michael D. 2001. Archaeofaunal relative abundances, sample size, and statistical methods. *Journal of Archaeological Science* 28: 185-195.

Clutton-Brock, Juliet 1987. *A Natural History of Domesticated Mammals*. Cambridge University Press, Cambridge.

Counihan, Carole M. 1999. *The Anthropology of Food and Body. Gender, Meaning, and Power*. Routledge, New York.

Coy, Jennie 1982. The role of wild vertebrate fauna in urban economies in Wessex. In Hall, A. R. &. Kenward, H. K (eds.). *Environmental Archaeology in the Urban Context*. The Council for British Archaeology, London 1982, p. 107-116.

Crosby, Alfred W. 1986. *Ecological Imperialism. The Ecological Expansion of Europe, 900-1900*. Cambridge University Press, Cambridge.

Damuth, John & MacFadden, Bruce J. (eds.) 1990. *Body Size in Mammalian Paleobiology. Estimation and Biological Implications.* Cambridge University Press, Cambridge.

Davis, S. J. M. 2000. The effect of castration and age on the development of the Shetland sheep skeleton and metric comparison between bones of males, females and castrates. *Journal of Archaeological Science* 27, 373–390.

Deagan, Kathleen A. 1996. Environmental archaeology and historical archaeology. Reitz, E. J., Newsom, L. A. & Scudder, S. J. (eds). *Case Studies in Environmental Archaeology.* New York, Plenum Press, 359-376.

Descola, Philippe & Pálsson, Gísli (eds.) 1996. *Nature and Society. Anthropological Perspectives.* Routledge, London.

Dinzelbacher, Peter 2002. Animal trials: a multidisciplinary approach. J*ournal of Interdisciplinary History* XXXII:3, 405-421.

von den Driesch, Angela 1976. *Das Vermessen von Tierknochen aus Vor- und Frühgeschichtlichen Siedlungen.* Univeristät München, München.

von den Driesch, Angela & Boessneck, Joachim 1974. Kritische Anmerkungen zur Widerristhöhenberecnhung aus Längenmaßen vor- und frühgeschichlicher Tierknochen. *Säugetierkundliche Mitteilungen* 4/1974.

Douglas, Mary 1984. *Purity and Danger. An Analysis of the Concepts of Pollution and Taboo.* Routledge, London.

Douglas, Mary 1994. The pangolin revisited: a new approach to animal symbolism. In Willis (ed.) 1994, p. 25-36.

Douglas, Mary 1999. Animals in Lele religious symbolism. *Implicit Meanings. Essays in Cultural Meaning.* Routledge, London, 34-46.

During, Ebba 1986. *The Fauna of Alvastra. An Osteological Analysis of Animal Bones from a Neolithic Pile Dwelling.* Stockholm Studies in Archaeology 6.

Ellen, Roy & Fukui, Katsuyoshi (eds.). 1996. *Redefining Nature. Ecology, Culture and Domestication.* Berg, Oxford.

Falk, Ann-Britt 2008. *En grundläggande handling: byggnadsoffer och dagligt liv i medeltid.* Nordic Academic Press, Lund.

Fiore, D. & Zangrando, A.F.J. 2006. Painted Fish, Eaten Fish. Artistic and Archaeofaunal Representations in Tierra del Fuego, Southern South America. *Journal of Anthropological Archaeology* 25(3), 371-389.

Fisher, John W. Jr. 1995. Bone surface modifications in zooarchaeology. *Journal of Archaeological Method and Theory* 1:7-68.

Fjellström, Phebe 1993. Nainen Tornionlaaksossa. *Tornionlaakson historia II. 1600-luvulta vuoteen 1809.* Tornionlaakson kuntien historiatoimikunta, Haapranta. p. 320-337.

Franklin, Adrian 1999. *Animals and Modern Cultures. A Sociology of Human-Animal Relations in Modernity.* Sage Publications, London.

Grant, Annie 1982. The use of tooth wear as a guide to the age of domestic ungulates. In Wilson, Bob, Grigson, Caroline & Payne, Sebastian (eds). *Ageing and Sexing Animal Bones from Archaeological Sites.* BAR British Series 109, 91-108.

Grayson, Donald K. 1994. *Quantitative Zooarchaeology.* Academic Press, Orlando.

Greenfield, Haskel J. 2006. Sexing fragmentary ungulate acetabulae. Ruscillo, Deborah (ed.). *Recent Advances in Ageing and Sexing Animal Bones.* Oxbow Books, Oxford. p. 68-86.

Grigson, Caroline 1982. Sex and age determination of some bones and teeth of domestic cattle: a review of the literature. In Wilson, Bob, Grigson, Caroline & Payne, Sebastian (eds). *Ageing and Sexing Animal Bones from Archaeological Sites.* BAR British Series 109, 7-24.

Grotenfelt, Gösta 1916. *Vanhanaikainen suomalainen maitotalous.* Otava, Helsinki.

Haapanen, Eero 1999. *Menneisyyden Helsingin eläimet. Pääkaupungin nisäkkäät, matelijat ja sammakkoeläimet arkistolähteissä vuosina 1850-1980.* Helsingin kaupungin ympäristökeskus, Helsinki.

Halila, Aimo 1954. *Pohjois-Pohjanmaan ja Lapin historia V. Pohjois-Pohjanmaa ja Lappi 1721-1775.* Kirjola, Oulu.

Hastorf, Christine A. & Johannessen, Sissel 1996. Understanding changing people/plant relationships in the Prehispanic Andes. Preucel, Robert & Hodder, Ian (eds.). *Contemporary Archaeology in Theory.* Blackwell, Oxford.

Hell, Bertrand 1996. Enraged hunters: the domain of the wild in north-western Europe. In Descola & Pálsson (eds) 1996, p. 205-218.

Herva, Vesa-Pekka 2003. *Tornio Keskikatu 29-35. Kaupunkiarkeologinen pelastuskaivaus.* Unpublished excavation report. University of Oulu Laboratory of Archaeology, Oulu.

Herva, Vesa-Pekka, forthcoming. Building persons: relationality and the life of buildings in an early modern Swedish town. *Antiquity*.

Herva, Vesa-Pekka & Ylimaunu, Timo 2005. Posliiniastiat, varallisuus ja kuluttajakäyttäytyminen 1700-luvun Torniossa. *Suomen Museo* 112, 79-89.

Herva, Vesa-Pekka & Ylimaunu, Timo 2009. Folk beliefs, special deposits and engagement with the environment in early modern northern Finland. *Journal of Anthropological Archaeology* 28(2): 234-243.

Herva, Vesa-Pekka & Nurmi, Risto 2009. Beyond consumption: functionality, object biography and early modernity in a European periphery. *International Jounal of Historical Archaeology* 13(2):158-182.

Herva, Vesa-Pekka, Ylimaunu, Timo, Kallio-Seppä, Titta, Kuokkanen, Tiina & Nurmi, Risto forthcoming. Maps, plots and curtains: material culture and the making of boundaries in an early modern town.

Higham, C.F.W. 1969. The metrical attributes of two samples of bovine limb bones. Journal of Zoology 157:63-74.

Holm, Ingunn 2002. A cultural landscape beyond the infield/outfield categories: an example from Eastern Norway. *Norwegian Archaeological Review* 35(2), pp. 67-80.

Howell, Signe 1996. Nature in culture or culture in nature? Chewong ideas of 'humans' and other species. In Descola & Pálsson (eds.) 1996, p.127-144.

Hufthammer, Anne Karin 1995. Age determination of reindeer (Rangifer tarandus L.). *Archaeozoologia* 7(2): 33-42.

Hukantaival, Sonja 2007. Hares's feet under a hearth – Discussing 'ritual' deposits in buildings. Immonen, Visa, Lempiäinen, Mia & Rosendahl, Ulrika (eds.). *Hortus Novus. Fresh Approaches to Medieval Archaeology in Finland.* Suomen keskiajan arkeologian seura, Turku. p. 66-75.

Hänninen, Eira 2008. *Oulun Virastotalon tontin vuoden 2007 kaivausten yksikön BSY 25/35 eläinluulöydöt.* Unpublished Master's Thesis. University of Oulu, Oulu.

Hållans, Ann-Mari & Andersson, Carolina 1992. Acquiring, using and discarding – Consumption patterns in the seventeenth century town of Nyköping. L. Ersgård, M. Holmström & K. Lamm (eds), *Rescue and Research – Reflections of Society in Sweden 700-1700* A.D. Riksantikvarieämbetet, Stockholm, 191-220.

Ijzereef, F. Gerard 1989. Social differentiation from animal bone studies. In D. Serjeantson & T. Waldron (eds.), *Diet and Crafts in Towns. The Evidence of Animal Remains from the Roman to the Post-Medieval Periods*. BAR, British Series 199, 41-53.

Ingold, Tim 2000. *Perception of the Environment: Essays on Livelihood, Dwelling and Skill.* Routledge, London.

Ingold, Tim 2006. Rethinking the animate, re-animating thought. *Ethnos* 71(1): 9-20.

Jarva, Eero, Niskanen, Markku & Paavola, Kirsti 2001. Anatomy of a late Iron Age inhumation burial of Hiukka at Nivankylä (Rovaniemi, Finnish Lapland). *Fennoscandia Archaeologica* XVIII: 27-49.

Jolley, Robert 1983. North American historic sites zooarchaeology. *Historical Archaeology* 17(2), 64-79.

Jones, Andrew K. G. 1982. Bulk-sieving and the recovery of fish remains from urban archaeological sites. Hall, A. R. & Kenward., H. K. (eds.). *Environmental Archaeology in the Urban Context.* Research Report no 43. The Council for British Archaeology, London, p. 79-85.

Jones, Andrew 1998. Where eagles dare. Landscape, animals and the Neolithic of Orkney. *Journal of Material Culture* 3(3).

Jones, Siân 1999. Historical categories and the praxis of identity: the interpretation of ethnicity in historical archaeology. Funari, Pedro P., Hall, Martin, Jones, Siân (eds.) *Historical Archaeology : Back from the Edge.* Routledge, London. p. 219-232.

Keskisarja, Teemu 2006. *"Secoituxesta järjettömäin luondocappalden canssa" Perversiot, oikeuselämä ja kansankulttuuri 1700-luvun Suomessa.* Helsingin yliopisto, Helsinki.

Klein, Richard G. & Cruz-Uribe, Kathryn 1984. *The Analysis of Animal Bones from Archaeological Sites.* The University of Chicago Press, Chicago.

Klemettinen, Pasi 2002. Kurkistuksia karhun kulttuurihistoriaan. Ilomäki, Henni & Lauhakangas, Outi (eds). *Eläin ihmisen mielenmaisemassa.* Suomalaisen Kirjallisuuden Seura, Helsinki.

Knight, John 1996. When timber grows wild: the desocialisation of Japanese mountain forests. In Descola & Pálsson (eds.) 1996, p. 221-239.

Krebs, Charles J. 1999. *Ecological Methodology.* 2nd edition. Addison, Wesley, Longman, Menlo Park.

Kvist, Roger 1987. *Sälfångsten i Österbotten och Västerbotten 1551-1610. En översikt och problemställning*, Umeå Universtity, Umeå.

Kvist, Roger 1990. *Sälfångstens roll in den lokala ekonomin. Österbotten och Västerbotten 1551-1610.* Center for Arctic Cultural Research, Umeå Universtity.

Lahti, Eeva-Kristiina 2006. Bones from Sápmi: reconstruction of the everyday life of two ancient Saami households. In: Herva, V. (Ed.), *People, Material Culture and Environment in the North*. Proceedings of the 22nd Nordic Archaeological Conference. University of Oulu, Oulu, 284-295.

Lam, Y.M., Xingbin, Chen, & Pearson, O.M. 1999. Intertaxonomic variability in patterns of bone density and the differential representation of Bovid, Cervid, and Equid elements in the archaeological record. *American Antiquity* 64(2):343-362.

Landon, David B. 2005. Zooarchaeology and historical archaeology: progress and prospects. *Journal of Archaeological Method and Theory* 12(1), 1-36.

Landon, David B. 1996. Feeding Colonial Boston: A Zooarchaeological Study. *Historical Archaeology* 30.

Lokki, Juhani & Palmgren, Jörgen 1992. *Suomen ja Pohjolan linnut.* 3. painos. WSOY, Helsinki.

Lukkarinen, J. 1939. Lehmän- ja härännimiä 1700-luvulta. *Kalevalaseuran vuosikirja* 19:197-211.

Luukko, Armas 1954. *Pohjois-Pohjanmaan ja Lapin historia II. Pohjois-Pohjanmaan ja Lapin keskiaika sekä 1500-luku.* Pohjois-Pohjanmaan maakuntaliiton ja Lapin maakuntaliiton yhteinen historiatoimikunta, Oulu.

Lyman, R. Lee 1994. *Vertebrate Taphonomy.* Cambridge Manuals in Archaeology. Cambridge University Press, Cambridge.

Lyman, R. Lee 2008. *Quantitative Palaeozoology.* Cambridge University Press, Cambridge.

Lyman, R. Lee 2006. Identifying bilateral pairs of deer (*Odocoileus* sp.) bones: how symmetrical is symmetrical enough? *Journal of Archaeological Science* 33: 1256-1265.

Magurran, Anne E. 2004. *Measuring Biological Diversity.* Blackwell, Malden.

Maltby, Mark 1979. *The Animal Bones from Exeter 1971-1975.* Exeter Archaeological Reports 2, University of Sheffield, Sheffield.

Mannermaa, Kristiina 2008a. Birds and burials at Ajvide (Gotland, Sweden) and Zvejnieki (Latvia) about 8000-3900 BP. *Journal of Anthropological Archaeology* 27(2):201-225.

Mannermaa, Kristiina 2008b. *The Archaeology of Wings. Birds and People in the Baltic Sea Region During the Stone Age.* PhD Dissertation, University of Helsinki.

Marciniak, Arkadiusz 1999. Faunal materials and interpretive archaeology – epistemology reconsidered. *Journal of Archaeological Method and Theory* 6(4), 293-318.

Marciniak, Arkadiusz 2005. *Placing Animals in the Neolithic. Social Zooarchaeology of Prehistoric Farming Communities.* UCL Press, London.

Martin, Calvin 1978. *Keepers of the Game. Indian-Animal Relationships and the Fur Trade.* University of California Press, Berkeley.

Massa, Ilmo 1994. *Pohjoinen luonnonvalloitus. Suunnistus ympäristöhistoriaan Lapissa ja Suomessa.* Gaudeamus, Helsinki.

Mayer, John J. & Brisbin, Lehr Jr, 1988. Sex identification of Sus scrofa based on canine morphology. *Journal of Mammalogy* 69(2), 408-412.

Morris, James T. 2005. Red deer's role in social expression on the Isles of Scotland. Pluskowski, Aleksander (ed.). *Just Skin and Bones? New Perspectives on Human-Animal Relations in the Historical Past.* BAR International Series 1410. BAR Publishing, Oxford. pp. 9-18.

Mrozovski, Stephen A. 1999. Colonization and the commodification of nature. *International Journal of Historical Archaeology* 3:3, 153-166.

Mäntylä, Ilkka 1971. *Tornion kaupungin historia. 1. osa. 1620-1809.* Tornion kaupunki, Tornio.

Nieminen, Mauri 1994. *Poro. Ruumiinrakenne ja elintoiminnat.* Riista- ja kalatalouden tutkimuskeskus, Rovaniemi.

Nieminen, Mauri & Helle, Timo 1980. Variation in body measurements of wild and semi-domestic reindeer (*Rangifer tarandus*) in Fennoscandia. *Annales Zoologici Fennici* 17: 275–283.

Nieminen, Mauri & Petersson, Carl Johan 1990. Growth and relationship of live weight to body measurements in semi-domesticated reindeer (*Rangifer tarandus tarandus* L.). *Rangifer Special Issue* 3: 353–361.

Niskanen, Markku 2006. Stature of the Merovingian-period inhabitants from Leväluhta, Finland. *Fennoscandia Archaeologica* 2006:24-36.

Niskanen, Markku 2008. A shift in animal species used for food from the Early Iron Age to the Roman Period. Försen, B. (ed.). *Thesprotia Expedition I, Papers and Monographs of the Finnish Institute at Athens XIV*. Finnish Institute at Athens, Helsinki. pp. 61-70.

Noddle, Barbara A. 1971. Determination of the body weight of cattle from bone measurements. Matolcsi, János (ed). *Domestikationsforschung und Geschichte der Haustiere*. Akadémiai Kiadó, Budapest, 377-389.

Nummi, Petri (ed.) 1997. S*uomen luonto. Nisäkkäät*. 3rd edition. WSOY, Helsinki.

Nurmi, Risto 2004. *Ab urbe Torna condita. Varallisuuden ilmeneminen Tornion kaupungin varhaisvaiheessa kahden kesällä 2002 tutkitun rakennuksen vertailun perusteella*. Unpublished Master's thesis. University of Oulu, Oulu.

Nurmi, Risto 2005a. TANK-02/RYNKS-02 *Kaupunkiarkeologinen pelastuskaivaus, Tornio, Keskikatu 29-35. Stratigrafisten yksiköiden ajoitukset*. Unpublished dating report. University of Oulu Laboratory of Archaeology, Oulu.

Nurmi, Risto 2005b. TKR-04 T*ornio, Välikatu 19. Omakotitalon perustusten maanvaihtotöiden valvonta ja säilyneiden 1700-luvun kerrosten rakenteiden dokumentointi*. Unpublished excavation report. National Board of Antiquities, Department of Monuments and Sites.

O'Connor, Terry 2000. *The Archaeology of Animal Bones*. Texas A&M University Press, s.l.

O'Connor, T.P. 2003. *The Analysis of Urban Animal Bone Assemblages: A Handbook for Archaeologists*. Council for British Archaeology, York.

Outram, Alan K. 1999. A comparison of Paleo-Eskimo and Medieval Norse bone fat exploitation in Western Greenland. *Arctic Anthropology* 36(1-2):103-117.

Outram, Alan K. 2001. A new approach to identifying bone marrow and grease exploitation: why the "indeterminate" fragments should not be ignored. *Journal of Archaeological Science* 28: 401–410.

Pálsson, Gísli 1994. The idea of fish: land and sea in the Icelandic world-view. In Willis (ed.) 1994, 119-133.

Pluskowski, Aleksander 2005. Prowlers in dark and wild places: mapping wolves in Medieval Britain and Southern Scandinavia. Pluskowski, Aleksander (ed.). *Just Skin and Bones? New Perspectives on Human-Animal Relations in the Historical Past*. BAR International Series 1410. BAR Publishing, Oxford. pp. 81-94.

Price, Cynthia R. 1985. Patterns of cultural behaviour and intra-site distributions of faunal remains at the Widow Harris site. *Historical Archaeology* 19(2): 40-56.

Puputti, Anna-Kaisa 2005. Eläinluututkimuksia 1600-luvun Oulusta – Kajaaninkadun ja Byströmin talon luulöydöt. Kallio, T. & Lipponen, S. (eds.), *Historiaa kaupungin alla – Kaupunkiarkeologisia tutkimuksia Oulussa*. Pohjois-Pohjanmaan museo, Oulu: 77-83.

Puputti, Anna-Kaisa 2006a. A zooarchaeological study on animal husbandry and game exploitation in seventeenth century Tornio, Northern Finland. *META* 1/2006: 13-28.

Puputti, Anna-Kaisa 2006b. Bones, economic strategies and socioeconomic status: an analysis of two bone assemblages from seventeenth century Tornio. *Fennoscandia Archaeologica* 2006:47-54.

Puputti, Anna-Kaisa 2007. Ruuanjätteitä tunkiolla. Pikisaaren kesän 2006 kaivausten eläinluulöydöt. *Faravid* 31: 9-22.

Puputti, Anna-Kaisa 2008. A zooarchaeology of modernizing human-animal relationships in Tornio, northern Finland, 1620-1800. *Post-Medieval Archaeology* 42(2): 304-316.

Puputti, Anna-Kaisa 2009a. Mitä tekemistä eläinosteologialla on arkeologisen tutkimuksen kanssa? *Muinaistutkija* 1/2009: xx-xx.

Puputti, Anna-Kaisa 2009b. P*ietarsaaren Lassfolkin 2008 kaivausten eläinten luut. Eläinosteologinen raportti*. Unpublished report. National Board of Antiquities, Department of Monuments and Sites.

Puputti, Anna-Kaisa in press a. Bones in pits and ditches. A contextual approach to animal bone distribution in early modern Tornio. *Journal of Nordic Archaeological Science*.

Puputti, Anna-Kaisa in press b. Sealing in Northern Finnish coastal towns in the seventeenth and eighteenth centuries: A zooarchaeological perspective. *The Second International Colloquium of Fishery, Trade, Piracy – Baltic and North Sea in the Middle Ages and later conference paper*.

Puputti, Anna-Kaisa & Niskanen, Markku 2008. The estimation of body weight of the reindeer *(Rangifer tarandus* L.) from skeletal Measurements: preliminary analyses and application to archaeological material from seventeenth and eighteenth century Northern Finland. *Environmental Archaeology* 13(2):153-164.

Puputti, Anna-Kaisa & Niskanen, Markku 2009. Identification of semi-domesticated reindeer (*Rangifer tarandus tarandus*, Linnaeus 1758) and wild forest reindeer (*R.t.fennicus*, Lönnberg 1909) from postcranial skeletal measurements. Mammalian Biology 74(1): 304-316.

Pääkkönen, Mirva 2006. *Tornion Keskikadun kesän 2002 kaivausten nuoremman punasavikeramiikan tarjoilu- ja säilytysastiat.* Unpublished Master's thesis. University of Oulu, Oulu.

Ranta, Raimo 1981. Suurvalta-ajan kaupunkilaitos. In Tommila, P., (ed.), S*uomen kaupunkilaitoksen historia 1. Keskiajalta 1870-luvulle.* Suomen kaupunkiliitto, Helsinki. pp. 7-134.

Reitz, Elizabeth & Wing, Elizabeth 2004. *Zooarchaeology.* 5th edition. Cambridge University Press, Cambridge.

Rosén, Christina. 1999. *Föremål och social status i Halmstad ca 1550-1750.* Göteborgs Universitet, Göteborg.

Saladin D'Anglure, Bernard 1994. Nanook, super-male: the polar bear in the imaginary space and social time of the Inuit of the Canadian Arctic. In Willis (ed.) 1994.

Salisbury, Joyce E. 1994. T*he Beast Within. Animals in the Middle Ages.* Routledge, New York.

Salo, Eveliina 2007. *Liitupiippujen kuluttaminen Torniossa 1600-1700-luvuilla: Keskikadun vuoden 2002 kaivausten liitupiippuaineiston käyttöjälkitutkimus.* Unpublished Master's thesis. University of Oulu, Oulu.

Sax, Boria 2000. *Animals in the Third Reich. Pets, Scapegoats, and the Holocaust.* Continuum, New York.

Schanche, Audhild 2002. Meahcci – den samiske utmarka. Andersen, Svanhild (ed.). *Samiske landskap og Agenda 21. Kultur, næring, miljøvern og demokrati.* Dieđut 1/2002. Sámi Instituhtta, Kautokeino.

Scholliers, Peter 2001. Meals, food narratives, and sentiments of belonging in past and present. Scholliers, Peter (ed.), *Food, Drink and Identity. Cooking, Eating and Drinking in Europe since the Middle Ages.* Berg, Oxford. p. 3-22.

Scott, Kathleen M. 1990. Postcranial dimensions of ungulates as predictors of body mass. Damuth, John & MacFadden, Bruce J. (eds.) 1990. *Body Size in Mammalian Paleobiology. Estimation and Biological Implications.* Cambridge University Press, Cambridge, 301-335.

Screen, J.E.O. 2007. *The Army in Finland During the Last Decades of Swedish Rule (1770-1809).* Suomalaisen Kirjallisuuden Seura, Helsinki.

Serjeantson, Dale 2000. Good to eat and good to think with: classifying animals from complex sites. In Peter Rowley-Conwy (ed.), *Animal Bones, Human Societies.* Oxbow Books, Oxford, 179-189.

Shaffer, Brian S. 1992. Quarter-inch screening: Understanding biases in recovery of vertebrate faunal remains. American Antiquity 57(1): 129-136.

Silver, I. A. 1969. The ageing of domestic animals. In D. Brothwell & E. Higgs (eds). *Science in Archaeology. A Survey of Progress and Research. 2nd edition.* Thames and Hudson, London, 283-302.

Soderberg, John 2004. Wild Cattle: Red Deer in the Religious Texts, Iconography, and Archaeology of Early Medieval Ireland. *International Journal of Historical Archaeology* 8 (3).

Soininen, Arvo M. 1974. *Vanha maataloutemme. Maatalous ja maatalousväestö Suomessa perinnäisen maatalouden loppukaudella 1720-luvulta 1870-luvulle.* Suomen maataloustieteellinen seura, Helsinki.

Stallibrass, Sue 2000. Dead dogs, dead horses: site formation processes at Ribchester Roman fort. In Rowley-Conwy, P. (ed.). *Animal Bones, Human Societies.* Oxbow Books, Oxford. p. 158-165.

Stark-Arola, Laura 2001. Women and food in rural-traditional Finland. Social and symbolic dimensions. *Elore* 2/2001.

Sten, Sabine 1994. Storleksvariationer hos medeltida och nyare tids nötkreatur och får. Svenska husdjur från medeltid till våra dagar. Skrifter om skogs- och lantbrukshistoria 5. Nordiska Museet, Stockholm, 35-50.

Sten, Sabine 2004. *Bovine Teeth in Age Assessment, from Medieval Cattle to Belgian Blue. Methodology, Possibilities and Limitations.* Stockholms universitet, Stockholm.

Storå, Jan 2001. *Reading Bones. Stone Age Hunters and Seals in the Baltic.* University of Stockholm, Stockholm.

Svensson, Emma M., Götherström, Anders & Vretemark, Maria 2008. A DNA test for sex identification in cattle confirms osteometric results. *Journal of Archaeological Science* 35(4): 942-946.

Sykes, Naomi 2005. Hunting for the Anglo-Normans: zooarchaeological evidence for Medieval identity. Pluskowski, Aleksander (ed.). *Just Skin and Bones? New Perspectives on Human-Animal Relations in the Historical Past.* BAR International Series 1410. BAR Publishing, Oxford. pp. 73-80.

Tagliacozzo, Antonio & Gala, Monica 2001. Exploitation of Anseriformes at two Upper Palaeolithic sites in Southern Italy: Grotta Romanelli (Lecce, Apulia) and Grotta del Santuario della Madonna a Praia a Mare (Cosenza, Calabria). *Acta Zoologica Cracoviensia* 45: 117-131.

Talve, Ilmar 1996. Suomenlahden ulkosaarten kansankulttuuri 1800-luvun loppupuolelta talvisotaan. In Hamari, Risto, Korhonen, Martti, Miettinen, Timo & Talve, Ilmar (eds.). *Suomenlahden ulkosaaret. Lavansaari, Seiskari, Suursaari, Tytärsaari*. Suomalaisen Kirjallisuuden Seura, Helsinki. p. 117-253.

Talve, Ilmar 1973. *Suomen kansanomaisesta ruokataloudesta*. Turun yliopiston kansatieteenlaitoksen toimituksia 2.

Talve, Ilmar 1990. *Suomen Kansankulttuuri*. 3rd edition. Suomalaisen kirjallisuuden seura, Helsinki.

Tapper, Richard L. 1988. Animality, humanity, morality, society. Ingold, Tim (ed.). *What is an Animal?* Unwin Hyman, London. p. 47-62.

Tarkka, Lotte 1994. Metsolan merkki – metsän olento ja kuva vienalaisrunostossa. Laaksonen, P. and Mettomäki, S., (eds.), *Metsä ja metsänviljaa*. Kalevalaseuran vuosikirja 73. pp. 56-102.

Tarkka, Lotte 1998. Sense of the forest: nature and gender in Karelian oral poetry. Apo, Satu, Nenola, Aili & Stark-Arola, Laura (eds.). *Gender and Folklore. Perspectives on Finnish and Carelian Culture*. Suomalaisen Kirjallisuuden Seura, Helsinki. p. 92-142.

Tarlow, Sarah 2007. *The Archaeology of Improvement in Britain, 1750-1850*. Cambridge University Press, Cambridge.

Tarlow, Sarah & West, Susie (eds.) 1999. *The Familiar Past? Archaeologies of Later Historical Britain*. Routledge, London.

Teerijoki, Ilkka 1993. Maatalouselinkeinot ja maatalouspolitiikka. *Tornionlaakson historia II. 1600-luvulta vuoteen 1809*. Tornionlaakson kuntien historiatoimikunta, Haapranta. p. 70-97.

Thomas, Julian 1998. *Time, Culture and Identity: An Interpretive Archaeology*. Routledge, London.

Thomas, Keith 1984. *Man and the Natural World. Changing Attitudes in England 1500-1800*. Penguin Books, Harmondsworth.

Thomas, Richard 2005. Zooarchaeology, improvement and the British agricultural revolution. *International Journal of Historical Archaeology* 9(2), 71-88.

Tourunen, Auli 2003. Eläinten luita kaupunkikerroksista-esimerkkejä arkeo-osteologisista tutkimusmenetelmistä. In Seppänen, Liisa (ed). *Kaupunkia pintaa syvemmältä – Arkeologisia näkökulmia Turun historiaan*. Suomen Keskiajan Arkeologian Seura, Turku, 371-382.

Tourunen, Auli 2008. *Animals in an Urban Context. A Zooarchaeological study of the Medieval and Post-Medieval town of Turku*. Turun yliopiston julkaisuja. Humaniora. Turun yliopisto, Turku.

Uerpmann, Hans-Peter 1973. Animal bone finds and economic archaeology: a critical study of osteo-archaeological method. *World Archaeology* 4(3): 307-322.

Ukkonen, Pirkko 2002. The early history of seals in the northern Baltic. *Annales zoologici fennici* 39(3): 187-207.

Vahtola, Jouko 1997. Vaikea vuosisata (1601-1721). Satokangas, Reija (ed.). *Keminmaan historia*. Keminmaan kunta, Keminmaa. p. 86-177.

Vainio-Korhonen, Kirsi 1999. Mamselli Falckin tarina eli kaupunkihistorian tutkimaton sukupuoli. *Ihmiset ovat kaupunki*. Turun Historiallinen Yhdistys, Turku. p. 65-78.

Whittle, Alasdair 2003. *The Archaeology of People. Dimensions of Neolithic Life*. Routledge, London.

Vilkuna, A. 1976. Kansanomainen karjasuoja. Lannanhuollolliset tyypit. Vuorela, Toivo (ed.). *Suomen kansankulttuurin kartasto 1. Aineellinen kulttuuri*. SKS, Helsinki.

Villa, Paola & Mahieu, Eric 1990. Breakage patterns of human long bones. *Journal of Human Evolution* 21: 27-48.

Willerslev, Rane 2001. The hunter as a human "kind": hunting and shamanism among the Upper Kolyma Yukaghirs of Siberia. *North Atlantic Studies* 4: 44–50.

Willerslev, Rane 2007. *Soul Hunters. Hunting, Animism, and Personhood among the Siberian Yukaghirs*. University of California Press, Berkeley.

Willis, Roy (ed.) 1994. *Signifying Animals. Human Meaning in the Natural World*. Routledge, London.

Wilmi, Jorma 2003. Tuotantotekniikka ja ravinnonsaanti. In Rasila, Viljo, Jutikkala, Eino & Mäkelä-Alitalo, Anneli (eds.). *Suomen maatalouden historia I. Perinteisen maatalouden aika. Esihistoriasta 1870-luvulle*. Suomalaisen Kirjallisuuden Seura, Helsinki. p. 159-182

Virkkunen, A.H. 1953. *Oulun kaupungin historia I. Kaupungin alkuajoilta isonvihan loppuun, 1610-1721.* 2. painos. Oulun kaupunki, Oulu.

Virrankoski, Pentti 1973. *Pohjois-Pohjanmaan ja Lapin historia III. Pohjois-Pohjanmaa ja Lappi 1600-luvulla.* Pohjois-Pohjanmaan ja Lapin maakuntaliiton yhteinen historiatoimikunta, Oulu.

Viveiros de Castro, Edouard 1998. Cosmological deixis and Amerindian perspectivism. *The Journal of the Royal Anthropological Institute* 4(3), 469-488.

Vuorela, Toivo 1975. *Suomalainen Kansankulttuuri.* WSOY, Porvoo.

Vuorisalo, Timo & Virtanen, Tapio 1988. Mätäjärven luulöydöt. *Turun Mätäjärvi.* Turun Maakuntamuseon Raportteja 10, 222-229.

Vretemark, Maria 1997. *Från ben till boskap. Kosthåll och djurhållning med utgångspunkt i medeltida benmaterial från Skara.* Skrifter från Länsmuseet Skara 25. Skaraborgs länsmuseum, Skara.

Vretemark, Maria 2001. Om nyttan av nötdjur. Andrén, Anders, Ersgård, Lars & Wienberg, Jes (eds.). *Från stad till land. En medeltidsarkeologisk resa tillägnad Hans Andersson.* Almqvist & Wiksell International, Stockholm 2001. pp. 45-50.

Vretemark, Maria 2003. Om livsmedelsförsörjning och sophantering. Karlsson, Pär & Tagesson, Göran (eds.), *I Tyskebacken. Hus, människor och industri in stormaktstidens Norrköping.* Riksantikvarieämbetet, Linköping. p. 84-97.

Ylimaunu, Juha 1996. Hylkeenpyynti pohjoisella Perämerellä. *Tornionlaakson vuosikirja* 1996: 181-211.

Ylimaunu, Juha 2000. *Itämeren hylkeenpyyntikulttuurit ja ihminen-hylje-suhde.* Suomalaisen Kirjallisuuden seura, Helsinki.

Ylimaunu, Juha 2002. Elinkeinot ihmisen ja eläimen suhteen muokkaajina. Ilomäki, Henni & Lauhakangas, Outi (eds). *Eläin ihmisen mielenmaisemassa.* Suomalaisen Kirjallisuuden Seura, Helsinki.

Ylimaunu, Timo 1996. *Kertomus Tornion Aspion ja Viippolan tonttien kaupunkiarkeologisesta koetutkimuksesta kesällä 1996.* Unpublished excavation report. Tornionlaakson maakuntamuseo, Tornio.

Ylimaunu, Timo 2000. *Kertomus kaupunkiarkeologisista tutkimuksista Torniossa, Keskikatu 13 –tontilla vuonna 1999 (T13-99).* Unpublished excavation report. Tornionlaakson maakuntamuseo, Tornio.

Ylimaunu, Timo 2001. *Purran ja Ahon tonttien kaupunkiarkeologiset tutkimukset Torniossa vuonna 1999.* Unpublished excavation report. Tornionlaakson maakuntamuseo, Tornio.

Ylimaunu, Timo 2007. *Aittakylästä kaupungiksi – arkeologinen tutkimus Tornion kaupungistumisesta 18. vuosisadan loppuun mennessä.* Pohjois-Suomen Historiallinen Yhdistys, Rovaniemi.

Zachrisson, Inger & Iregren, Elisabeth 1974. *Lappish Bear Graves in Northern Sweden. An Archaeological and Osteological Study.* Almqvist & Wiksell, Stockholm.

Zar, Jerold H. 1996. *Biostatistical Analysis.* 3rd edition. Prentice Hall, New Jersey.

Åström, Sven-Erik 1978. *Natur och byte. Ekologiska synpunkter på Finlands ekonomiska historia.* Söderström, Helsinki.

Electronic sources

Web pages of the Animal Genetic Data Bank of the European Association for Animal Production. http://www.tiho-hannover.de/einricht/zucht/eaap/

Archive sources

OMA, Provincial Archives of Oulu
RA, Riksantikvarieämbetet

www.ingramcontent.com/pod-product-compliance
Ingram Content Group UK Ltd.
Pitfield, Milton Keynes, MK11 3LW, UK
UKHW061213180426
11947UKWH00029B/2018